SIMPLIFIED HIGHER ALGEBRA:

A SELF-TEACHING BOOK
FOR HIGH SCHOOLS AND COLLEGES

By

Kingsley Augustine

Table of Contents

1 SIMPLIFICATION OF ALGEBRAIC FRACTIONS 4
    Exercise 1 9

2 EQUATIONS AND SUBSTITUTIONS INVOLVING FRACTIONS 11
    Undefined Fractions 12
    Substitution in Algebraic Fractions 15
    Exercise 2 19

3 SIMULTANEOUS EQUATIONS INVOLVING FRACTIONS 21
    Exercise 3 27

4 ABSOLUTE VALUE EQUATION (MODULUS EQUATION) 28
    Exercise 4 40

5 INEQUALITIES INVOLVING ABSOLUTE VALUES, QUOTIENT AND SQUARE FUNCTIONS 42
    Inequalities Involving Quotients 45
    Inequalities Involving Square Functions 50
    Exercise 5 52

6 INDICIAL EQUATIONS 54
    Exercise 6 62

7 ROOTS OF QUADRATIC EQUATIONS (USE OF ALPHA AND BETA) 63
    Maximum and Minimum Values of a Quadratic Function 76
    Exercise 7 81

8 FUNCTIONS 84
    Domain and Range of a Function 86
    Arithmetic Operations of Function 87
    Composing Functions 90
    Continuous and Discontinuous Functions 93
    Even Functions 93
    Odd Functions 93
    Inverse of a Function 95
    Further Worked Examples on Functions 99
    Exercise 8 103

9 POLYNOMIALS 106

Addition and Subtraction of Polynomials 106

Multiplication of Polynomials... 107

Division of Polynomials 109

Zeros of Polynomials 115

The Factor Theorem... 116

The Remainder Theorem 116

Exercise 9... 125

10 PARTIAL FRACTION 127

Resolving Algebraic Fractions into Partial Fractions 127

Types of Partial Fraction 127

Exercise 10... 148

11 RADICAL EQUATIONS 150

Exercise 11... 157

SOLUTION TO EXERCISES... 158

CHAPTER 1
SIMPLIFICATION OF ALGEBRAIC FRACTIONS

When simplifying single algebraic fractions, factorize the numerator and denominator and then cancel out common terms. However, when two or more algebraic fractions are to be simplified, we treat them like the usual way of dealing with fractions by taking LCM and following the necessary steps.

Examples

1. Simplify $\dfrac{a^2 + ab}{a^2 + ac}$

<u>Solution</u>

Factorize the fraction to give:
$$\frac{a^2 + ab}{a^2 + ac} = \frac{a(a + b)}{a(a + c)}$$
The 'a' will cancel out to give the final answer as follows:
$$\frac{\cancel{a}(a + b)}{\cancel{a}(a + c)} = \frac{a + b}{a + c}$$
Note that we can only cancel out terms linking one another by multiplication. Terms that are linked together by addition sign or subtraction sign cannot be cancelled out. For example, in example 1 above, the fraction $\dfrac{a(a + b)}{a(a + c)}$ can also be written as: $\dfrac{a \times (a + b)}{a \times (a + c)}$. The 'a's linked by multiplication sign can be cancelled out as shown in our solution above. However, in the final answer given by $\dfrac{a + b}{a + c}$, the a's cannot be cancelled out to give $\dfrac{b}{c}$ since they are linked by addition sign.

2. Simplify $\dfrac{c^2 - 2c - 15}{c^2 - 3c - 10}$

<u>Solution</u>

The numerator and denominator are quadratic expressions. We factorize them as follows:
$$\frac{c^2 - 2c - 15}{c^2 - 3c - 10} = \frac{(c - 5)(c + 3)}{(c - 5)(c + 2)}$$
Therefore, (c – 5) will cancel out to give our final answer as follows:
$$\frac{\cancel{(c - 5)}(c + 3)}{\cancel{(c - 5)}(c + 2)} = \frac{(c + 3)}{(c + 2)}$$
Refer to my book 'Simplified Basic Algebra' for explanations on factorization of quadratic expression in order to understand how the factorization above was carried out.

3. Simplify $\dfrac{8 - 2a - a^2}{2a^2 - 3a - 2}$

Solution

$$\dfrac{8 - 2a - a^2}{2a^2 - 3a - 2}$$

Let us factorize $8 - 2a - a^2$ as follows:

Multiply the first and last terms to give:

$8(-a^2) = -8a^2$.

Two numbers in 'a' whose product is $-8a^2$ and sum is $-2a$ are $-4a$ and $2a$. Replace $-2a$ in the original expression with these two terms. This gives:

$8 - 4a + 2a - a^2$

We now factorize by grouping as follows:

$8 - 4a + 2a - a^2 = 4(2 - a) + a(2 - a)$

$\qquad\qquad\qquad\qquad = (2 - a)(4 + a)$ (After using $(2 - a)$ as a common term of the expression)

Similarly, we factorize the denominator as follows:

$2a^2 - 3a - 2 = 2a^2 - 4a + a - 2$

$\qquad\qquad\quad = 2a(a - 2) + 1(a - 2)$

$\qquad\qquad\quad = (a - 2)(2a + 1)$

$\therefore\ \dfrac{8 - 2a - a^2}{2a^2 - 3a - 2} = \dfrac{(2-a)(4+a)}{(a - 2)(2a + 1)}$

$\qquad\qquad\qquad = \dfrac{-(a - 2)(4+a)}{(a - 2)(2a + 1)}$ [Note that $(2 - a)$ can also be expressed as $-(a - 2)$]

Therefore, $(a - 2)$ will cancel out to give:

$\dfrac{-(a - 2)(4+a)}{(a - 2)(2a + 1)} = \dfrac{-\cancel{(a - 2)}(4 + a)}{\cancel{(a - 2)}(2a + 1)}$

$\qquad\qquad\qquad = \dfrac{-(4 + a)}{(2a + 1)}$

Note that an expression such as $(a - 2)$ can be converted to $-(2 - a)$. This is done by simply putting a negative sign outside the bracket and changing the signs of the two terms inside the bracket. Similarly, $(5 - x)$ can also be changed to $-(x - 5)$. We only carry out this kind of conversion in order to make it look the same like another term, so that the two terms can be canceled out during the simplification.

4. Simplify: $\dfrac{9a^2 - m^2}{m^2 - 2am - 3a^2}$

Solution

The numerator is a difference of two squares. Recall that a difference of two squares is factorized as

follows:
$$a^2 - b^2 = (a + b)(a - b)$$
Similarly, $9a^2 - m^2 = (3a)^2 - m^2$
$$= (3a + m)(3a - m)$$

In order to factorize the denominator, we multiply the first and last terms, since it is a quadratic expression. This gives:
$$m^2(-3a^2) = -3a^2m^2$$

Two numbers in 'am' whose product is $-3a^2m^2$ and sum is $-2am$ are $-3am$ and am.

$\therefore \quad m^2 - 2am - 3a^2 = m^2 - 3am + am - 3a^2$
$$= m(m - 3a) + a(m - 3a)$$
$$= (m - 3a)(m + a)$$

$\therefore \quad \dfrac{9a^2 - m^2}{m^2 - 2am - 3a^2} = \dfrac{(3a + m)(3a - m)}{(m - 3a)(m + a)}$

$$= \dfrac{-(3a + m)(m - 3a)}{(m - 3a)(m + a)}$$ (Note that $(3a - m) = -(m - 3a)$, as shown)

Hence, $(m - 3a)$ cancels out as follows:

$$\dfrac{-(3a + m)(m - 3a)}{(m - 3a)(m + a)} = \dfrac{-(3a + m)\cancel{(m - 3a)}}{\cancel{(m - 3a)}(m + a)}$$

$$= \dfrac{-(3a + m)}{(m + a)}$$

5. Simplify: $\dfrac{a^2 - am - an + mn}{a^2 - am + an - mn}$

<u>Solution</u>

Factorize both numerator and denominator by grouping terms. This gives:

$$\dfrac{a^2 - am - an + mn}{a^2 - am + an - mn} = \dfrac{a(a - m) - n(a - m)}{a(a - m) + n(a - m)}$$

$$= \dfrac{(a - m)(a - n)}{(a - m)(a + n)}$$

Therefore, $(a - m)$ cancels out as follows:

$$\dfrac{(a - m)(a - n)}{(a - m)(a + n)} = \dfrac{\cancel{(a - m)}(a - n)}{\cancel{(a - m)}(a + n)}$$

$$= \dfrac{(a - n)}{(a + n)}$$

6. Simplify $\dfrac{a+2}{a} - \dfrac{1}{3ab}$

Solution

$\dfrac{a+2}{a} - \dfrac{1}{3ab}$

The LCM of a and 3ab is 3ab. Hence divide the LCM by each of the denominator and multiply the value obtained by the respective numerator. This means that 3ab ÷ a = 3b. Then multiply 3b by a + 2 (the numerator). Similarly, 3ab ÷ 3ab = 1. Then multiply 1 by 1(the numerator). Finally divide these multiplied terms by the LCM. This is as shown below.

$$\dfrac{a+2}{a} - \dfrac{1}{3ab} = \dfrac{3b(a+2) - 1(1)}{3ab}$$

$$= \dfrac{3ab + 6b - 1}{3ab}$$

7. Simplify: $\dfrac{1}{4x-2y} - \dfrac{1}{y-2x}$

Solution

$\dfrac{1}{4x-2y} - \dfrac{1}{y-2x}$

A careful look at one of the denominator (i.e. 4x – 2y), shows that it can be factorized and made to look like the other denominator. This is carried out as follows.

$$\dfrac{1}{4x-2y} - \dfrac{1}{y-2x} = \dfrac{1}{2(2x-y)} - \dfrac{1}{y-2x}$$

$$= \dfrac{1}{2(2x-y)} - \dfrac{1}{-(2x-y)} \qquad \text{[Note that (y – 2x) is also –(2x – y)]}$$

$$= \dfrac{1}{2(2x-y)} + \dfrac{1}{(2x-y)} \qquad \text{(Note that –(–) has become +)}$$

The LCM of 2(2x – y) and (2x – y) is 2(2x – y). Hence we continue our simplification as follows:

$$\dfrac{1}{2(2x-y)} + \dfrac{1}{(2x-y)} = \dfrac{1+2}{2(2x-y)}$$

$$= \dfrac{3}{2(2x-y)}$$

8. Simplify $\dfrac{3}{2ab} + \dfrac{4}{3bc}$

Solution

$$\frac{3}{2ab} + \frac{4}{3bc}$$

The LCM of 2ab and 3bc is 6abc. Therefore:

$$\frac{3}{2ab} + \frac{4}{3bc} = \frac{3c(3) + 2a(4)}{6abc}$$

$$= \frac{9c + 8a}{6abc}$$

9. Simplify $\dfrac{3x}{x-1} - \dfrac{4}{x+2}$

Solution

$$\frac{3x}{x-1} - \frac{4}{x+2}$$

The LCM of $(x-1)$ and $(x+2)$ is $(x-1)(x+2)$. We now simplify as follows:

$$\frac{3x}{x-1} - \frac{4}{x+2} = \frac{(x+2)3x - 4(x-1)}{(x-1)(x+2)}$$

$$= \frac{3x^2 + 6x - 4x + 4}{(x-1)(x+2)}$$

$$= \frac{3x^2 + 2x + 4}{(x-1)(x+2)}$$

10. Simplify $\dfrac{1}{3xy} + \dfrac{5}{4y^2z} - \dfrac{3y}{2x^3}$

Solution

$$\frac{1}{3xy} + \frac{5}{4y^2z} - \frac{3y}{2x^3}$$

The LCM of $3xy$, $4y^2z$ and $2x^3$ is $12x^3y^2z$. Hence, we use this LCM to simplify the expression as follows:

$$\frac{1}{3xy} + \frac{5}{4y^2z} - \frac{3y}{2x^3} = \frac{1(4x^2yz) + 5(3x^3) - 3y(6y^2z)}{12x^3y^2z}$$

$$= \frac{4x^2yz + 15x^3 - 18y^3z}{12x^3y^2z}$$

11. Simplify $\dfrac{3y}{y^2 - z^2} - \dfrac{3z}{z^2 - y^2}$

Solution

$$\frac{3y}{y^2-z^2} - \frac{3z}{z^2-y^2}$$

The denominators are difference of two squares. Let us factorize the denominators and make them look alike.

$$\frac{3y}{y^2-z^2} - \frac{3z}{z^2-y^2} = \frac{3y}{(y-z)(y+z)} - \frac{3z}{(z-y)(z+y)}$$

$$= \frac{3y}{(y-z)(y+z)} - \frac{3z}{-(y-z)(y+z)}$$

Note that (z − y) is also −(y − z). Note also that (z + y) is the same as (y + z) as rearranged above. Hence the −(−) will become + as follows:

$$\frac{3y}{(y-z)(y+z)} + \frac{3z}{(y-z)(y+z)}$$

The LCM of (y − z)(y + z) and (y − z)(y + z) is simply (y − z)(y + z). When terms are the same, just take one of them as the LCM. Hence, we now continue with the simplification as follows:

$$\frac{3y}{(y-z)(y+z)} + \frac{3z}{(y-z)(y+z)} = \frac{3y(1)+3z(1)}{(y-z)(y+z)} \quad \text{(Note that } (y-z)(y+z) \div (y-z)(y+z) = 1)$$

$$= \frac{3y+3z}{(y-z)(y+z)}$$

$$= \frac{3(y+z)}{(y-z)(y+z)}$$

$$= \frac{3\cancel{(y+z)}}{(y-z)\cancel{(y+z)}}$$

$$= \frac{3}{(y-z)}$$

Exercise 1

1. Simplify $\dfrac{m^2+md}{m^2+mf}$

2. Simplify $\dfrac{x^2-4x-21}{x^2-x-12}$

3. Simplify $\dfrac{6-x-x^2}{3x^2+4x-15}$

4. Simplify $\dfrac{16m^2 - n^2}{n^2 - 3mn - 4m^2}$

5. Simplify: $\dfrac{p^2 - pq - pr + qr}{p^2 - pq + pr - qr}$

6. Simplify $\dfrac{2a + 3}{b} - \dfrac{2}{3b}$

7. Simplify: $\dfrac{3}{10x - 5y} - \dfrac{2}{y - 2x}$

8. Simplify $\dfrac{1}{6mn} + \dfrac{3}{2n}$

9. Simplify $\dfrac{5a}{a - 2} - \dfrac{2}{a + 3}$

10. Simplify $\dfrac{2}{8ab} + \dfrac{1}{3a^2 b} - \dfrac{3a}{12b^3}$

11. Simplify $\dfrac{4}{9m^2 - 4n^2} - \dfrac{5n}{4n^2 - 9m^2}$

CHAPTER 2
EQUATIONS AND SUBSTITUTIONS INVOLVING FRACTIONS

Examples

1. Solve the equation $\dfrac{3}{2b-5} - \dfrac{4}{b-3} = 0$

<u>Solution</u>

$\dfrac{3}{2b-5} - \dfrac{4}{b-3} = 0$

Multiply each term by the LCM, i.e. $(2b-5)(b-3)$. This gives:

$(2b-5)(b-3)\dfrac{3}{2b-5} - (2b-5)(b-3)\dfrac{4}{b-3} = (2b-5)(b-3)0$

Therefore, we cancel the common denominators as follows:

$\cancel{(2b-5)}(b-3)\dfrac{3}{\cancel{2b-5}} - (2b-5)\cancel{(b-3)}\dfrac{4}{\cancel{b-3}} = 0$ (Note that zero multiplies a number to give zero)

$(b-3)3 - (2b-5)4 = 0$

$3b - 9 - 8b + 20 = 0$ (Take note of the negative sign outside the second bracket)

$3b - 8b = 9 - 20$

$-5b = -11$

$b = \dfrac{-11}{-5}$

$b = 2\dfrac{1}{5}$ (Note that the negative sign has cancelled out)

2. Solve the equation $\dfrac{a-4}{7} = \dfrac{2}{3a-1}$

<u>Solution</u>

$\dfrac{a-4}{7} = \dfrac{2}{3a-1}$

We can multiply each term in the equation by the LCM, i.e. $7(3a-1)$. However, since there is a single fraction on each side of the equation, a simple way to solve this kind of equation is to cross multiply. This gives:

$(a-4)(3a-1) = 7(2)$

$3a^2 - a - 12a + 4 = 14$

$3a^2 - 13a + 4 - 14 = 0$

$3a^2 - 13a - 10 = 0$

Solving this quadratic equation by factorization gives:

$3a^2 - 15a + 2a - 10 = 0$

Note that $-15a$ and $+2a$ are the two numbers whose product will give $-30a^2$ (i.e. $3a^2 \times (-10) = -30a^2$)

and whose sum will give −13a. Hence they are used to substitute for −13a in the original quadratic equation. Therefore, we factorize by grouping as follows:
$$3a(a-5) + 2(a-5) = 0$$
$$(a-5)(3a+2) = 0$$
$$\therefore \quad a = 5 \text{ or } a = -\frac{2}{3}$$

3. Solve the equation $\dfrac{3}{x-4} = \dfrac{2}{x-1} - 4$

Solution
$$\frac{3}{x-4} = \frac{2}{x-1} - 4$$
Multiply each term by the LCM, i.e. $(x-4)(x-1)$. This gives:
$$(x-4)(x-1)\frac{3}{x-4} = (x-4)(x-1)\frac{2}{x-1} - (x-4)(x-1)4$$
We now cancel out the common terms as follows:
$$\cancel{(x-4)}(x-1)\frac{3}{\cancel{x-4}} = (x-4)\cancel{(x-1)}\frac{2}{\cancel{x-1}} - (x-4)(x-1)4$$

$$(x-1)3 = (x-4)2 - 4(x-4)(x-1)$$
$$3x - 3 = 2x - 8 - 4(x^2 - x - 4x + 4)$$
$$3x - 3 = 2x - 8 - 4x^2 + 4x + 16x - 16$$
$$3x - 3 = 22x - 24 - 4x^2$$
$$4x^2 + 3x - 22x - 3 + 24 = 0$$
$$4x^2 - 19x + 21 = 0$$

Solving this quadratic equation by factorization gives:
$$4x^2 - 12x - 7x + 21 = 0$$
$$4x(x-3) - 7(x-3) = 0$$
$$(x-3)(4x-7) = 0$$
$$x = 3 \text{ or } x = \frac{7}{4}$$
$$x = 3 \text{ or } x = 1\frac{3}{4}$$

Undefined Fractions
A fraction is undefined if the denominator of the fraction is zero.

Examples
1. Find the value of x for which the fraction $\dfrac{8}{15 + 3x}$ is undefined.

Solution

$$\frac{8}{15+3x}$$

For this fraction to be undefined, the denominator must be equal to zero. Hence:

$15 + 3x = 0$ (Note that only the denominator is equated to zero)

$3x = -15$

$x = \frac{-15}{3}$

$x = -5$

∴ The fraction is undefined when $x = -5$

2. Find the value of x for which the fraction $\frac{6x}{2x-5}$ is not defined.

Solution

For this fraction not to be defined, the denominator must be equal to zero. Hence:

$2x - 5 = 0$

$2x = 5$

$x = \frac{5}{2}$

$x = 2\frac{1}{2}$

∴ The fraction is not defined when $x = 2\frac{1}{2}$

3. For what value(s) of b is the fraction $\frac{2b+11}{b^2+b-20}$

a. not defined
b. equal to zero?

Solution

a. $\frac{2b+11}{b^2+b-20}$

For this fraction not to be defined, we equate the denominator to zero as follows:

$b^2 + b - 20 = 0$

Solving this quadratic equation by factorization gives:

$(b + 5)(b - 4) = 0$

Therefore, b = −5 or b = 4

Hence, the expression is not defined when b = −5 or b = 4

b. Recall that a fraction is zero when the numerator of the fraction is zero. Therefore we now equate the numerator to zero as follows:

$$2b + 11 = 0$$
$$2b = -11$$
$$b = \frac{-11}{2}$$
$$b = -5\frac{1}{2}$$

The fraction is equal to zero when $b = -5\frac{1}{2}$

4. For what value(s) of x is the expression $\frac{x^2+3x+2}{2x+3}$
   a. undefined
   b. zero?

Solution

a. $\frac{x^2+3x+2}{2x+3}$

For this expression to be undefined:
$$2x + 3 = 0$$
$$2x = -3$$
$$x = \frac{-3}{2}$$

Therefore, $x = -1\frac{1}{2}$

Hence, the expression is undefined when $x = -1\frac{1}{2}$

b. This expression is zero if:
$$x^2 + 3x + 2 = 0$$

Solving this quadratic equation by factorization gives:
$$x^2 + 3x + 2 = 0$$
$$(x+1)(x+2) = 0$$
$$x = -1 \text{ or } x = -2$$

∴ The expression is zero when $x = -1$ or $x = -2$

5. For what values of x is the expression $\frac{2x^2+3x-2}{3x^2-14x+8}$
   a. zero
   b. undefined?

Solution

a. $\dfrac{2x^2 + 3x - 2}{3x^2 - 14x + 8}$

This expression is zero if the numerator is zero. Therefore, we equate it to zero as follows:

$2x^2 + 3x - 2 = 0$

Let us solve this quadratic equation by factorization as follows:

$2x^2 + 3x - 2 = 0$
$2x^2 + 4x - x - 2 = 0$
$2x(x + 2) - 1(x + 2) = 0$
$(x + 2)(2x - 1) = 0$
$x = -2$ or $x = \dfrac{1}{2}$ (After equating each bracket to zero and solving for x)

∴ The expression is zero when $x = -2$ or $x = \dfrac{1}{2}$

b. The expression is undefined if the denominator is zero. Therefore, we equate it to zero as follows:

$3x^2 - 14x + 8 = 0$

Let us solve this quadratic equation by factorization as follows:

$3x^2 - 14x + 8 = 0$
$3x^2 - 12x - 2x + 8 = 0$
$3x(x - 4) - 2(x - 4) = 0$
$(x - 4)(3x - 2) = 0$
$x = 4$ or $x = \dfrac{2}{3}$ (After equating each bracket to zero and solving for x)

∴ The expression is undefined when $x = 4$ or $x = \dfrac{2}{3}$

Substitution in Algebraic Fractions

Examples

1. If $y = \dfrac{a^2 - b}{c - b}$, find the value of y when a = −2, b = −1 and c = 4

Solution

$y = \dfrac{a^2 - b}{c - b}$

We simply substitute the various given values of a, b and c into the equation in order to obtain the value of y. This gives:

$y = \dfrac{a^2 - b}{c - b}$

$$= \frac{(-2)^2 - (-1)}{4 - (-1)}$$

$$= \frac{4+1}{4+1}$$

$$= \frac{5}{5}$$

$$y = 1$$

2. If $x = \frac{m^3 + 5n - p}{2mp + m - 3n^2}$, find the value of x when m = –1, n = 3 and p = 6

Solution

$$x = \frac{m^3 + 5n - p}{2mp + m - 3n^2}$$

We simply substitute the various given values of m, n and p into the equation in order to obtain the value of x. This gives:

$$x = \frac{m^3 + 5n - p}{2mp + m - 3n^2}$$

$$= \frac{(-1)^3 + 5(3) - 6}{2(-1)(6) + (-1) - 3(3)^2}$$

$$= \frac{-1 + 15 - 6}{-12 - 1 - 27}$$

$$= \frac{8}{-40}$$

$$x = -\frac{1}{5}$$

3. If $p : q = 9 : 5$, evaluate $\frac{15p - 2q}{5p + 16q}$

Solution

An easy way of solving this problem is to substitute 9 for p and 5 for q in the given expression.

$$\therefore \frac{15p - 2q}{5p + 16q} = \frac{15(9) - 2(5)}{5(9) + 16(5)}$$

$$= \frac{135 - 10}{45 + 80}$$

$$= \frac{125}{125}$$

$$= 1$$

4. If $a : b = 5 : 3$, evaluate $\dfrac{6a + b}{a - \frac{1}{3}b}$

Solution

An easy way of solving this problem is to substitute 5 for a and 3 for b in the given expression.

$$\therefore \frac{6a + b}{a - \frac{1}{3}b} = \frac{6(5) + 3}{5 - \frac{1}{3}(3)}$$

$$= \frac{30 + 3}{5 - 1}$$

$$= \frac{33}{4}$$

$$= 8\frac{1}{4}$$

5. If $\dfrac{m}{n} = \dfrac{3}{4}$, evaluate $\dfrac{7m + n}{2m - \frac{1}{5}n}$

Solution

An easy way of solving this problem is to substitute 3 for m and 4 for q in the given expression.

$$\therefore \frac{7m + n}{2m - \frac{1}{5}n} = \frac{7(3) + 4}{2(3) - \frac{1}{5}(4)}$$

$$= \frac{21 + 4}{6 - \frac{4}{5}}$$

$$= \frac{25}{\frac{30 - 4}{5}}$$

$$= \frac{25}{\frac{26}{5}}$$

$$= \frac{25}{1} \times \frac{5}{26}$$

$$= \frac{125}{26}$$

$$= 4\frac{21}{26}$$

6. If $a = \frac{x+2}{x-1}$, express $\frac{2a+3}{a-1}$ in terms of x

Solution

In order to express $\frac{2a+3}{a-1}$ in terms of x, we simply substitute the expression for 'a' in $\frac{2a+3}{a-1}$. Since 'a' is expressed in terms x, then $\frac{2a+3}{a-1}$ will become expressed in x if 'a' is substituted in it. Therefore, let us substitute $\frac{x+2}{x-1}$ for 'a' in $\frac{2a+3}{a-1}$. This gives:

$$\frac{2a+3}{a-1} = \frac{2\left(\frac{x+2}{x-1}\right)+3}{\frac{x+2}{x-1}-1}$$

$$= \frac{\left(\frac{2x+4}{x-1}\right)+3}{\frac{x+2}{x-1}-1}$$

$$= \frac{\left(\frac{2x+4+3(x-1)}{x-1}\right)}{\frac{x+2-1(x-1)}{x-1}}$$

$$= \frac{\left(\frac{2x+4+3x-3}{x-1}\right)}{\frac{x+2-x+1)}{x-1}}$$

$$= \frac{\left(\frac{5x+1}{x-1}\right)}{\frac{3}{x-1}}$$

$$= \frac{5x+1}{x-1} \times \frac{x-1}{3}$$

$$= \frac{5x+1}{\cancel{x-1}} \times \frac{\cancel{x-1}}{3}$$

$$= \frac{5x+1}{3}$$

7. If $m = \dfrac{y+5}{y-3}$, express $\dfrac{5m+2}{m-3}$ in terms of y

Solution

In order to express $\dfrac{5m+2}{m-3}$ in terms of y, we simply substitute the expression for m in $\dfrac{5m+2}{m-3}$. Since m is expressed in terms y, then $\dfrac{5m+2}{m-3}$ will become expressed in y if m is substituted in it. Therefore, let us substitute $\dfrac{y+5}{y-3}$ for m in $\dfrac{5m+2}{m-3}$. This gives:

$$\dfrac{5m+2}{m-3} = \dfrac{5\left(\dfrac{y+5}{y-3}\right) + 2}{\left(\dfrac{y+5}{y-3}\right) - 3}$$

$$= \dfrac{\left(\dfrac{5y+25}{y-3}\right) + 2}{\left(\dfrac{y+5}{y-3}\right) - 3}$$

$$= \dfrac{\left(\dfrac{5y+25+2(y-3)}{y-3}\right)}{\dfrac{y+5-3(y-3)}{y-3}}$$

$$= \dfrac{\left(\dfrac{5y+25+2y-6}{y-3}\right)}{\dfrac{y+5-3y+9)}{y-3}}$$

$$= \dfrac{\left(\dfrac{7y+19}{y-3}\right)}{\dfrac{14-2y}{y-3}}$$

$$= \dfrac{7y+19}{y-3} \times \dfrac{y-3}{14-2y}$$

$$= \dfrac{7y+19}{\cancel{y-3}} \times \dfrac{\cancel{y-3}}{14-2y}$$

$$= \dfrac{7y+19}{14-2y}$$

Exercise 2
1. Solve the equation $\dfrac{2}{m-2} - \dfrac{3}{m-5} = 0$

19

2. Solve the equation $\dfrac{b-1}{5} = \dfrac{5}{2b-7}$

3. Solve the equation $\dfrac{1}{x-5} = \dfrac{9}{2x-12} - 4$

4. Find the value of x for which the fraction $\dfrac{1}{8+x}$ is undefined.

5. Find the value of x for which the fraction $\dfrac{2x}{3x-7}$ is not defined.

6. For what value(s) of a is the fraction $\dfrac{5a+12}{a^2-a-30}$
   a. zero
   b. undefined?

7. For what value(s) of m is the expression $\dfrac{2m^2+8m+8}{5m+9}$
   a. zero
   b. undefined?

8. For what values of x is the expression $\dfrac{5x^2+13x-6}{2x^2-x-15}$
   a. zero
   b. undefined?

9. If $m = \dfrac{p^2-q}{q-r}$, find the value of m when p = –5, q = 3 and r = -2

10. If $x = \dfrac{2a^3+5b-8c}{7ab+b-3c^2}$, find the value of x when a = –2, b = 5 and c = - 4

11. If m : n = 2 : 7, evaluate $\dfrac{6m+4n}{12m-2n}$

12. If a : b = 9 : 5, evaluate $\dfrac{2a+3b}{3a-\frac{1}{5}b}$

CHAPTER 3
SIMULTANEOUS EQUATIONS INVOLVING FRACTIONS

Examples

1. Solve the simultaneous equation: $\dfrac{x}{2} - \dfrac{y}{3} = \dfrac{1}{6}$

    $\dfrac{x}{2} - \dfrac{y}{6} = 5$

<u>Solution</u>

$\dfrac{x}{2} - \dfrac{y}{3} = \dfrac{1}{6}$ ....Equation (1)

$\dfrac{x}{2} - \dfrac{y}{6} = 5$ ....Equation (2)

This is a case where the variable is in the numerator. In cases like this, we clear out the fractions by multiplying each term by the LCM of the denominators. The LCM in equation (1) and (2) is 6. Hence we clear out the fractions by multiplying each term in equation (1) and (2) by 6. This gives:

$6(\dfrac{x}{2}) - 6(\dfrac{y}{3}) = 6(\dfrac{1}{6})$ ....Equation (1)

$6(\dfrac{x}{2}) - 6(\dfrac{y}{6}) = 6(5)$ ....Equation (2)

These now simplify to give equation 3 and equation 4 respectively, and we solve them simultaneously as shown below:

$3x - 2y = 1$ ....Equation (3)

$3x - y = 30$ ....Equation (4)

Equation (4) – equation (3): $y = 29$ (Note that –y –(–2y) = –y + 2y = y. Also, 30 – 1 = 29)

In equation (3), substitute 29 for y in order to obtain x. This gives:

$3x - 2y = 1$ ....Equation (3)

$3x - 2(29) = 1$

$3x - 58 = 1$

$3x = 1 + 58$

$3x = 59$

$x = \dfrac{59}{3}$

Hence, $x = \dfrac{59}{3}$ and y = 29

2. Solve simultaneously: $\dfrac{1}{x} + \dfrac{1}{y} = 5$

$\dfrac{1}{y} - \dfrac{1}{x} = 1$

Solution

METHOD 1

$\dfrac{1}{x} + \dfrac{1}{y} = 5$

$\dfrac{1}{y} - \dfrac{1}{x} = 1$

In these equations the variables are in the denominators. So, let us take $\dfrac{1}{x} = a$ and $\dfrac{1}{y} = b$. Hence we can rewrite the equations as follows:

a + b = 5

b − a = 1

Or,

$\qquad$ a + b = 5 ………………….Equation (1)

$\qquad$ −a + b = 1 ………………Equation (2)

Equation (1) + equation (2):   2b = 6

$\qquad\qquad\qquad$ b = $\dfrac{6}{2}$

$\qquad\qquad\qquad$ b = 3

Substitute 3 for b in equation (1). This gives:

$\qquad$ a + b = 5 ………………….Equation (1)

$\qquad$ a + 3 = 5

$\qquad$ a = 5 − 3

$\qquad$ a = 2

Hence a = 2 and b = 3

But recall that $\dfrac{1}{x} = a$ and $\dfrac{1}{y} = b$, and we are actually solving for x and y. Therefore, we now substitute the values of a and b appropriately, in order to obtain x and y as follows:

$\dfrac{1}{x} = a$

$\dfrac{1}{x} = 2$ (Since a = 2)

When we cross multiply it gives:

$\qquad$ 2x = 1

$\qquad$ x = $\dfrac{1}{2}$

Similarly, $\dfrac{1}{y} = b$

$\frac{1}{y} = 3$ (Since b = 3)

When we cross multiply it gives:

$3y = 1$

$y = \frac{1}{3}$

Therefore, $x = \frac{1}{2}$ and $y = \frac{1}{3}$

METHOD 2

$\frac{1}{x} + \frac{1}{y} = 5$

$\frac{1}{y} - \frac{1}{x} = 1$

Rearranging the equations gives:

$\frac{1}{x} + \frac{1}{y} = 5$ ....Equation (1)

$-\frac{1}{x} + \frac{1}{y} = 1$ ....Equation (2)

Since the coefficients of $\frac{1}{x}$ are the same in the two equations, but with different signs, then $\frac{1}{x}$ can be eliminated by using elimination method as follows:

$\frac{1}{x} + \frac{1}{y} = 5$ ....Equation (1)

$-\frac{1}{x} + \frac{1}{y} = 1$ ....Equation (2)

Equation (1) + equation (2): $\frac{2}{y} = 6$ (Note that $\frac{1}{y} + \frac{1}{y} = \frac{2}{y}$ and 5 + 1 = 6)

Hence, $y = \frac{2}{6}$

$y = \frac{1}{3}$

Substitute $\frac{1}{3}$ for y in equation 2 in order to obtain x. This gives:

$-\frac{1}{x} + \frac{1}{y} = 1$ ....Equation (2)

$-\frac{1}{x} + \frac{1}{\frac{1}{3}} = 1$

$$3 - 1 = \frac{1}{x}$$

$$2 = \frac{1}{x}$$

$$x = \frac{1}{2}$$

Hence, $x = \frac{1}{2}$ and $y = \frac{1}{3}$

3. Solve the simultaneous equations: $\frac{3}{c} - \frac{4}{d} = \frac{1}{3}$

$\frac{2}{c} - \frac{5}{d} = 1$

Solution

$$\frac{3}{c} - \frac{4}{d} = \frac{1}{3} \quad \text{..................Equation (1)}$$

$$\frac{2}{c} - \frac{5}{d} = 1 \quad \text{..................Equation (2)}$$

Let us make the coefficient of $\frac{1}{c}$ to be equal in the two equations so that it can be eliminated.

Therefore, multiply equation (1) by 2 and equation (2) by 3. This gives:

$$\frac{6}{c} - \frac{8}{d} = \frac{2}{3} \quad \text{..................Equation (3)}$$

$$\frac{6}{c} - \frac{15}{d} = 3 \quad \text{..................Equation (4)}$$

Equation (3) – Equation (4): $\quad \frac{7}{d} = -\frac{7}{3}$

Hence, $d = \frac{7 \times 3}{-7}$

$d = -3$

Substitute –3 for d in equation (2) in order to obtain c. This gives:

$$\frac{2}{c} - \frac{5}{d} = 1 \quad \text{..................Equation (2)}$$

$$\frac{2}{c} - \frac{5}{-3} = 1$$

$$\frac{2}{c} + \frac{5}{3} = 1$$

$$\frac{2}{c} = 1 - \frac{5}{3}$$

$$\frac{2}{c} = \frac{-2}{3}$$

24

$$c = \frac{2 \times 3}{-2}$$
$$c = -3$$
Therefore, $c = -3$ and $d = -3$

4. Solve the simultaneous equations: $\frac{a}{2} - \frac{b}{5} = 1$

$$b - \frac{a}{3} = 8$$

Solution

Rearranging the equations gives:

$\frac{a}{2} - \frac{b}{5} = 1$ ....Equation (1)

$-\frac{a}{3} + b = 8$ ....Equation (2)

Let us make the coefficient of a to be equal in the two equations so that it can be eliminated.

Therefore, multiply equation (1) by $\frac{1}{3}$ and equation (2) by $\frac{1}{2}$. This gives:

$\frac{a}{6} - \frac{b}{15} = \frac{1}{3}$ ....Equation (3)

$-\frac{a}{6} + \frac{b}{2} = 4$ ....Equation (4)

Equation (3) + Equation (4): $\frac{13b}{30} = \frac{13}{3}$

Hence, $b = \frac{30 \times 13}{13 \times 3}$

$b = 10$

Substitute 10 for b in equation (1) in order to obtain a. This gives:

$\frac{a}{2} - \frac{b}{5} = 1$ ....Equation (1)

$\frac{a}{2} - \frac{10}{5} = 1$

$\frac{a}{2} = 1 + 2$

$a = 3 \times 2$

$a = 6$

Hence, $a = 6$ and $b = 10$

5. Solve simultaneously the equations: $\dfrac{3}{m} + \dfrac{5}{n} = 1$

$\dfrac{9}{m} - \dfrac{5}{n} = 1$

Solution

$\dfrac{3}{m} + \dfrac{5}{n} = 1$ ………………..Equation(1)

$\dfrac{9}{m} - \dfrac{5}{n} = 1$ ………………..Equation(2)

Let us make the coefficient of $\dfrac{1}{m}$ to be equal in the two equations by multiplying equation (1) by 3. This gives:

$\dfrac{9}{m} + \dfrac{15}{n} = 3$ ………………..Equation (3)

$\dfrac{9}{m} - \dfrac{5}{n} = 1$ ………………..Equation (4)

Equation (3) – Equation (4): $\dfrac{20}{n} = 2$

Hence, $n = \dfrac{20}{2}$

$n = 10$

Substitute 10 for n in equation (1) in order to obtain m. This gives:

$\dfrac{3}{m} + \dfrac{5}{n} = 1$ ………………..Equation(1)

$\dfrac{3}{m} + \dfrac{5}{10} = 1$

$\dfrac{3}{m} = 1 - \dfrac{1}{2}$

$\dfrac{3}{m} = \dfrac{1}{2}$

$m = 3 \times 2$

$m = 6$

Hence, m = 6 and n = 10

Exercise 3

1. Solve the simultaneous equation: $\dfrac{x}{3} - \dfrac{y}{4} = -\dfrac{1}{6}$

   $\dfrac{x}{3} - \dfrac{y}{8} = \dfrac{1}{12}$

2. Solve simultaneously, the equation: $\dfrac{1}{a} + \dfrac{1}{b} = 3$

   $\dfrac{1}{a} - \dfrac{1}{b} = 5$

3. Solve the simultaneous equations: $\dfrac{2}{c} + \dfrac{3}{d} = 2$

   $\dfrac{3}{c} - \dfrac{1}{d} = 1\dfrac{1}{6}$

4. Solve the simultaneous equations: $\dfrac{a}{5} - \dfrac{b}{4} = \dfrac{1}{10}$

   $a - \dfrac{b}{2} = 1$

5. Solve simultaneously the equations: $\dfrac{4}{p} + \dfrac{6}{q} = 4$

   $\dfrac{1}{p} - \dfrac{3}{q} = -8$

6. Solve the simultaneous equations: $\dfrac{2}{c} - \dfrac{10}{d} = 3$

   $\dfrac{5}{c} - \dfrac{5}{d} = 3\dfrac{1}{2}$

7. Solve the simultaneous equations: $\dfrac{x}{2} - \dfrac{y}{4} = \dfrac{1}{20}$

   $\dfrac{x}{4} + \dfrac{y}{4} = \dfrac{7}{40}$

8. Solve simultaneously the equations: $\dfrac{1}{m} + \dfrac{1}{n} = 2$

   $\dfrac{3}{m} - \dfrac{6}{n} = 5$

CHAPTER 4
ABSOLUTE VALUE EQUATION (MODULUS EQUATION)

The absolute value of a number is a positive (or zero) value of the number. Whether the number was originally positive or negative, its absolute value must be positive. For example, $|7| = 7$ and $|-7| = 7$. This shows that for each absolute value of a number/expression, there are two possible numbers/expressions. Hence, when solving absolute value equations, we split the equation into two possible equations.

Examples
1. Solve $|x| = 5$

Solution
$|x| = 5$
Remove the absolute value bars and split the equation into two cases. The first case should have a positive right hand side, while the second case should have a negative right hand side. This gives:
$x = 5$ or $x = -5$

2. Solve the following equations:
a. $|3x - 2| = 4$
b. $|2x + 1| = 9$
c. $|7 - 5x| = 2$

Solutions
a. $|3x - 2| = 4$
Simply remove the bars and take a positive right hand value, and then a negative right hand value to obtain two possible equations. Then solve each of the equations separately. This gives:

$$3x - 2 = 4 \quad \text{or} \quad 3x - 2 = -4$$
$$3x = 4 + 2 \quad\quad\quad 3x = -4 + 2$$
$$3x = 6 \quad\quad\quad\quad 3x = -2$$
$$x = \frac{6}{3} \quad\quad\quad\quad x = -\frac{2}{3}$$
$$x = 2$$

Hence, $x = 2$ or $x = -\frac{2}{3}$

Take note of how the two equations were solved separately.

b. $|2x + 1| = 9$
Remove the bars and take a positive right hand value to form one equation. Form a second equation by also removing the bars and taking a negative right hand value. Then solve each of the equations

separately. This gives:

$2x + 1 = 9$ or $2x + 1 = -9$
$2x = 9 - 1$ $2x = -9 - 1$
$2x = 8$ $2x = -10$
$x = \dfrac{8}{2}$ $x = \dfrac{-10}{2}$
$x = 4$ $x = -5$

Hence, $x = 4$ or $x = -5$

c. $|7 - 5x| = 2$

$7 - 5x = 2$ or $7 - 5x = -2$
$7 - 2 = 5x$ $7 + 2 = 5x$
$5 = 5x$ $9 = 5x$
$x = \dfrac{5}{5}$ $5x = 9$
$x = 1$ $x = \dfrac{9}{5}$

Hence, $x = 1$ or $x = 1\dfrac{4}{5}$

3. Solve the following equation:
a. $|4x - 11| = 0$
b. $|2x - 5| = -1$

Solutions

a. $|4x - 11| = 0$

In this case there is only one equation that we can obtain. This is given by:

$4x - 11 = 0$
$4x = 11$
$x = \dfrac{11}{4}$

Hence, $x = 2\dfrac{3}{4}$

b. $|2x - 5| = -1$

Recall that an absolute value must be positive. But this equation is telling us that absolute value is negative. This is impossible. An absolute value cannot be negative (i.e. −1 in this case), hence there is no solution to this equation.

4. Solve the following equations:

a. $|x-5| + 8 = 12$
b. $|3x+7| - 14 = -4$

Solution

a. $|x-5| + 8 = 12$

Take the constant term (outside the bars) to the right hand side of the equation. This gives:

$|x-5| = 12 - 8$
$|x-5| = 4$

Let us now remove the bars and split this equation into two possible equations. This gives:

$x - 5 = 4$	or	$x - 5 = -4$
$x = 4 + 5$		$x = -4 + 5$
$x = 9$		$x = 1$

Hence, $x = 9$ or $x = 1$

b. $|3x+7| - 14 = -4$

Do not think that this equation has no solution since there is a negative value on the right hand side. If we simplify the equation we will get a positive value on the right hand side, with the absolute value bars still in place. Therefore, let us simplify the equation as follows.

Take the constant term to the right hand side of the equation. This gives:

$|3x+7| = -4 + 14$
$|3x+7| = 10$

Now that we have a positive value on the right hand side, let us remove the bars and split this equation into two possible equations. This gives:

$3x + 7 = 10$	or	$3x + 7 = -10$
$3x = 10 - 7$		$3x = -10 - 7$
$3x = 3$		$3x = -17$
$x = \dfrac{3}{3}$		$x = -\dfrac{17}{3}$
$x = 1$		$x = -5\dfrac{2}{3}$

Hence, $x = 1$ or $x = -5\dfrac{2}{3}$

5. Solve the following absolute value equations:
a. $|x-6| = 2x - 3$
b. $|5x+2| = x + 4$
c. $|3x-4| = |2x+3|$

Solution

a. $|x-6| = 2x-3$

Let us split the equation into two equations as follows:

$x-6 = 2x-3$	or	$x-6 = -(2x-3)$
$x-2x = -3+6$		$x-6 = -2x+3$
$-x = 3$		$x+2x = 3+6$
$x = -3$		$3x = 9$
		$x = \dfrac{9}{3}$
		$x = 3$

Hence, $x = -3$ or $x = 3$

Now, in equations such as this, it is necessary for us to test the values obtained to see if they are true solutions. In order to do this, we substitute the values of x into the right side of the equation (the side without the absolute value bars) to see if it will give us a positive value. A negative value means that the value of x obtained is not a solution to the equation. Let us write out the part of the equation without the absolute value bar as follows:

$2x-3$

When $x = -3$, the above expression gives:

$2x-3 = 2(-3) - 3$
$= -6 - 3 = -9$

Hence $x = -3$ is not a solution since it gives a negative right hand side. This is because an absolute value cannot be negative.

When $x = 3$, the above expression gives:

$2x-3 = 2(3) - 3$
$= 6 - 3 = 3$

Hence $x = 3$ is a solution since it gives a positive value on the right hand side.

Therefore the overall solution of the equation is $x = 3$

b. $|5x+2| = x+4$

Let us split the equation into two equations as follows:

$5x+2 = x+4$	or	$5x+2 = -(x+4)$
$5x-x = 4-2$		$5x+2 = -x-4$
$4x = 2$		$5x+x = -4-2$
$x = \dfrac{2}{4}$		$6x = -6$
$x = \dfrac{1}{2}$		$x = \dfrac{-6}{6}$
		$x = -1$

Let us now find out if these two values of x are solutions of the equations. In order to do this, we substitute the values of x into the right side of the equation (the side without the absolute value bars).

Let us write out the part of the equation without the absolute value bar as follows:

$x + 4$

When $x = \frac{1}{2}$, the above expression gives:

$$x + 4 = \frac{1}{2} + 4$$
$$= 4\frac{1}{2}$$

Hence $x = \frac{1}{2}$ is a solution since it gives a positive right hand side.

When $x = -1$, the above expression gives:

$$x + 4 = -1 + 4$$
$$= 3$$

Hence $x = -1$ is a solution since it gives a positive value on the right hand side.

Therefore the overall solutions of the equation are $x = \frac{1}{2}$ or $x = -1$

c. $|3x - 4| = |2x + 3|$

Let us split the equation into two equations as follows:

$3x - 4 = 2x + 3$	or	$3x - 4 = -(2x + 3)$
$3x - 2x = 3 + 4$		$3x - 4 = -2x - 3$
$x = 7$		$3x + 2x = -3 + 4$
		$5x = 1$
		$x = \frac{1}{5}$

Since this question has the absolute value bars on both side of the equation, then our two values of x are correct. There is no need to verify out solution.

Therefore, $x = 7$ or $x = \frac{1}{5}$

6. Solve the following absolute value equations:
a. $||5x + 2| - 5| = 12$
b. $||7 - 2x| - 8| = 5$

<u>Solution</u>

a. $||5x + 2| - 5| = 12$

This is a case where one absolute value expression is nested in another. Working with the outermost bars, let us split the overall equation into two equations. This gives:

$|5x + 2| - 5 = 12$	or	$|5x + 2| - 5 = -12$
$|5x + 2| = 12 + 5$		$|5x + 2| = -12 + 5$
$|5x + 2| = 17$		$|5x + 2| = -7$

Out of these two new equations that we have obtained, one of them has no solution. The equation $|5x + 2| = -7$ has no solution since an absolute value cannot be negative. Hence we discard $|5x + 2| = -7$. However, the other equation which is $|5x + 2| = 17$ can be solved. Hence, let us solve it by splitting it into two equations as follows:

$5x + 2 = 17$	or	$5x + 2 = -17$
$5x = 17 - 2$		$5x = -17 - 2$
$5x = 15$		$5x = -19$
$x = \dfrac{15}{5}$		$x = \dfrac{-19}{5}$
$x = 3$		$x = -3\dfrac{4}{5}$

Therefore, $x = 3$ or $x = -3\dfrac{4}{5}$

b. $||7 - 2x| - 8| = 5$

Starting with the outermost bars, we split the overall equation into two equations as follows:

$|7 - 2x| - 8 = 5$	or	$|7 - 2x| - 8 = -5$
$|7 - 2x| = 5 + 8$		$|7 - 2x| = -5 + 8$
$|7 - 2x| = 13$ ....Equation (1)		$|7 - 2x| = 3$ ....Equation (2)

Now we have to take each of the equations above and split it into two equations. Let us split equation (1) into two equations as follows:

$7 - 2x = 13$	or	$7 - 2x = -13$
$-2x = 13 - 7$		$-2x = -13 - 7$
$-2x = 6$		$-2x = -20$
$x = \dfrac{6}{-2}$		$x = \dfrac{-20}{-2}$
$x = -3$		$x = 10$

Hence $x = -3$ or $x = 10$

Let us take equation (2) above and split it into two equations as follows:

$7 - 2x = 3$	or	$7 - 2x = -3$
$-2x = 3 - 7$		$-2x = -3 - 7$
$-2x = -4$		$-2x = -10$
$x = \dfrac{-4}{-2}$		$x = \dfrac{-10}{-2}$
$x = 2$		$x = 5$

Hence $x = 2$ or $x = 5$

Therefore the overall solutions to the original equation are: $x = -3, 10, 2$ or 5

7. Solve the following equations:

a. $|2x-3| = |5x+1| + 1$
b. $|x+7| - |3x-1| = -4$

Solution

a. $|2x-3| = |5x+1| + 1$

There are three possible equations that we are going to solve.

(i) The first equation that we are going to solve is obtained by simply dropping the absolute value bars in the equation. This gives:

$2x - 3 = 5x + 1 + 1$
$2x - 5x = 1 + 1 + 3$
$-3x = 5$
$x = -\dfrac{5}{3}$

(ii) The second equation that we are going to solve is obtained by making the two absolute value terms to be negative. This gives:

$-(2x - 3) = -(5x + 1) + 1$
$-2x + 3 = -5x - 1 + 1$
$-2x + 5x = -1 + 1 - 3$
$3x = -3$
$x = \dfrac{-3}{3}$
$x = -1$

(iii) Lastly, the third equation that we have to solve is obtained by making only the absolute value term on the left hand side to be negative. This gives:

$-(2x - 3) = 5x + 1 + 1$
$-2x + 3 = 5x + 1 + 1$
$-2x - 5x = 1 + 1 - 3$
$-7x = -1$
$x = \dfrac{-1}{-7}$
$x = \dfrac{1}{7}$

Hence the three values of x obtained are $-\dfrac{5}{3}, -1$ and $\dfrac{1}{7}$.

Now we have to check which of these values are solutions of the equation. We do this by substituting each of the value of x into the equation. The equation is:

$|2x-3| = |5x+1| + 1$

When $x = -\dfrac{5}{3}$, we have:

$$|2(-\tfrac{5}{3})-3| = |5(-\tfrac{5}{3})+1|+1$$
$$|-\tfrac{10}{3}-3| = |-\tfrac{25}{3}+1|+1$$
$$|\tfrac{-10-9}{3}| = |\tfrac{-25+3}{3}|+1$$
$$|\tfrac{-19}{3}| = |\tfrac{-22}{3}|+1$$
$$\tfrac{19}{3} = \tfrac{22}{3}+1 \quad \text{(Take note of the removal of the negative sign when the bars are removed}$$
$$\tfrac{19}{3} = \tfrac{22+3}{3}$$
$$\tfrac{19}{3} = \tfrac{25}{3}$$

Since both sides of the equation are not equal, then $x = -\tfrac{5}{3}$ is not a solution of the equation.

When $x = -1$, we have:
$$|2x-3| = |5x+1|+1$$
$$|2(-1)-3| = |5(-1)+1|+1$$
$$|-2-3| = |-5+1|+1$$
$$|-5| = |-4|+1$$
$$5 = 4+1$$
$$5 = 5$$

Since both sides are equal, then $x = -1$ is a solution of the equation.

When $x = \tfrac{1}{7}$, we have:
$$|2(\tfrac{1}{7})-3| = |5(\tfrac{1}{7})+1|+1$$
$$|\tfrac{2}{7}-3| = |\tfrac{5}{7}+1|+1$$
$$|\tfrac{2-21}{7}| = |\tfrac{5+7}{7}|+1$$
$$|\tfrac{-19}{7}| = |\tfrac{12}{7}|+1$$
$$\tfrac{19}{7} = \tfrac{12}{7}+1$$
$$\tfrac{19}{7} = \tfrac{12+7}{7}$$
$$\tfrac{19}{7} = \tfrac{19}{7}$$

Since both sides are equal, then $x = \tfrac{1}{7}$ is a solution of the equation.

Therefore the solutions of the equation are $x = -1$ and $x = \tfrac{1}{7}$

b. $|x+7| - |3x-1| = -4$

Let us rearrange the equation in order to have only one of the absolute value expression on one side of the equation. This gives:

$|x+7| = |3x-1| - 4$

Let us solve each of the three possible equations from this equation.

(i) The first equation that we are going to solve is obtained by simply dropping the absolute value bars in the equation. This gives:

$x + 7 = 3x - 1 - 4$
$x - 3x = -1 - 4 - 7$
$-2x = -12$
$x = \dfrac{-12}{-2}$
$x = 6$

(ii) The second equation is obtained by making the two absolute value terms to be negative. This gives:

$-(x+7) = -(3x-1) - 4$
$-x - 7 = -3x + 1 - 4$
$-x + 3x = 1 - 4 + 7$
$2x = 4$
$x = \dfrac{4}{2}$
$x = 2$

(iii) The third equation that we have to solve is obtained by making only the absolute value term on the left hand side to be negative. This gives:

$-(x+7) = 3x - 1 - 4$
$-x - 7 = 3x - 1 - 4$
$-x - 3x = -1 - 4 + 7$
$-4x = 2$
$x = \dfrac{2}{-4}$
$x = -\dfrac{1}{2}$

Hence the three values of x obtained are 6, 2 and $-\dfrac{1}{2}$

Now, we have to check which of these values are solutions of the equation. We do this by substituting each of the values of x into the equation. The equation is:

$|x+7| = |3x-1| - 4$

When $x = 6$, we have:

$|6+7| = |3(6) - 1| - 4$
$|13| = |18 - 1| - 4$

$13 = |17| - 4$

$13 = 17 - 4$

$13 = 13$

Since both sides of the equation are equal, then $x = 6$ is a solution of the equation.

When $x = 2$, we have:

$|2 + 7| = |3(2) - 1| - 4$

$|9| = |6 - 1| - 4$

$9 = |5| - 4$

$9 = 5 - 4$

$9 = 1$

Since both sides of the equation are not equal, then $x = 2$ is not a solution of the equation.

When $x = -\frac{1}{2}$, we have:

$|-\frac{1}{2} + 7| = |3(-\frac{1}{2}) - 1| - 4$

$|\frac{-1+14}{2}| = |-\frac{3}{2} - 1| - 4$

$|\frac{13}{2}| = |\frac{-3-2}{2}| - 4$

$\frac{13}{2} = |\frac{-5}{2}| - 4$

$\frac{13}{2} = \frac{5}{2} - 4$

$\frac{13}{2} = \frac{5-8}{2}$

$\frac{13}{2} = -\frac{3}{2}$

Since both sides of the equation are not equal, then $x = -\frac{1}{2}$ is not a solution of the equation.

Therefore the only solution of the equation is $x = 6$

8. Solve the following equations:

a. $-2|4x + 9| = -22$

b. $3|x - 2| + 5 = -2|x - 2| + 10$

<u>Solution</u>

a. $-2|4x + 9| = -22$

Divide both sides by −2. This gives:

$|4x + 9| = \frac{-22}{-2}$

$|4x + 9| = 11$

We now split this equation into two equations as follows:

$$4x + 9 = 11 \quad \text{or} \quad 4x + 9 = -11$$
$$4x = 11 - 9 \qquad\qquad 4x = -11 - 9$$
$$4x = 2 \qquad\qquad 4x = -20$$
$$x = \frac{2}{4} \qquad\qquad x = \frac{-20}{4}$$
$$x = \frac{1}{2} \qquad\qquad x = -5$$

Therefore, $x = \frac{1}{2}$ or $x = -5$

b. $3|x-2| + 5 = -2|x-2| + 10$

This equation is similar to $3m + 5 = -2m + 10$, where m is $|x-2|$. Hence we can solve for m. However, we are not going to use m, we will solve the equation using $|x-2|$ as a variable. This is done as follows.

$$3|x-2| + 5 = -2|x-2| + 10$$

Collect terms in $|x-2|$ on the left hand side of the equation. This gives:

$3|x-2| + 2|x-2| = 10 - 5$ (The left hand side is similar to $3m + 2m$ which will give 5m)

$5|x-2| = 5$ (The left side is like 5m which is $5|x-2|$ if we imagine $m = |x-2|$)

Divide both sides by 5. This gives:

$$|x-2| = \frac{5}{5}$$
$$|x-2| = 1$$

We now split this equation into two equations as follows:

$$x - 2 = 1 \quad \text{or} \quad x - 2 = -1$$
$$x = 1 + 2 \qquad\qquad x = -1 + 2$$
$$x = 3 \qquad\qquad x = 1$$

Therefore, $x = 3$ or $x = 1$

9. Solve the following equations:

a. $|2x^2 - 7x + 6| = 0$

b. $|x^2 + 5x - 80| = 4$

<u>Solutions</u>

a. $|2x^2 - 7x + 6| = 0$

Since the right hand side of the equation is zero, we can split this equation to only one equation. This gives:

$$2x^2 - 7x + 6 = 0$$

Solving this quadratic equation by factorization gives:

$$2x^2 - 4x - 3x + 6 = 0$$
$$2x(x-2) - 3(x-2) = 0$$

$(x-2)(2x-3) = 0$

Hence, $x = 2$ or $\frac{3}{2}$.   (After equating each bracket above to zero and solving for x)

b. $|x^2 + 5x - 80| = 4$

We can split this equation into two equations as follows:

$x^2 + 5x - 80 = 4$ ....Equation (1)

$x^2 + 5x - 80 = -4$ ....Equation (2)

Let us solve equation (1) as follows:

$x^2 + 5x - 80 = 4$

$x^2 + 5x - 80 - 4 = 0$

$x^2 + 5x - 84 = 0$

Solving this equation by factorization gives:

$(x\quad)(x\quad) = 0$

We now find two numbers such that their product is –84 and their sum is +5. The two numbers are +12 and –7. The two number are entered into the brackets above to give:

$(x + 12)(x - 7) = 0$

Hence, $x = -12$ or $x = 7$

Let us now solve equation (2) above as follows:

$x^2 + 5x - 80 = -4$

$x^2 + 5x - 80 + 4 = 0$

$x^2 + 5x - 76 = 0$

Let us use quadratic formula to solve this equation. From the equation, a = 1, b = 5 and c = –76

$x = \dfrac{-b \pm \sqrt{b^2 - 4ac}}{2a}$

$= \dfrac{-5 \pm \sqrt{5^2 - [4(1)(-76)]}}{2(1)}$

$= \dfrac{-5 \pm \sqrt{25 - (-304)}}{2}$

$= \dfrac{-5 \pm \sqrt{25 + 304}}{2}$

$= \dfrac{-5 \pm \sqrt{329}}{2}$

$= \dfrac{-5 \pm 18.14}{2}$

$x = \dfrac{-5 + 18.14}{2}$ or $x = \dfrac{-5 - 18.14}{2}$

$$= \frac{13.14}{2} \quad \text{or} \quad \frac{-23.14}{2}$$

$$x = 6.57 \quad \text{or} \quad x = -11.57$$

Therefore, the four solutions of the equation are $x = -12, 7, 6.57$ and -11.57.

Exercise 4

1. Solve $|x| = 9$
2. Solve the following equations:
   a. $|5x + 7| = 17$
   b. $|3x + 1| = 8$
   c. $|15 - 4x| = 18$

3. Solve the following equation:
   a. $|7x + 13| = 8$
   b. $|5x - 27| = -14$

4. Solve the following equations:
   a. $|4x - 3| + 6 = 17$
   b. $|x + 9| - 11 = -7$

5. Solve the following absolute value equations:
   a. $|3x - 6| = x - 2$
   b. $|8x + 5| = 7x + 40$
   c. $|2x - 5| = |10x - 1|$

6. Solve the following absolute value equations:
   a. $||2x + 7| - 5| = 6$
   b. $||9 - 4x| - 8| = 1$

7. Solve the following equations:
   a. $|3x - 1| = |2x + 5| + 6$
   b. $|5x + 2| - |11x - 3| = -1$

8. Solve the following equations:
   a. $-5|4x + 9| = 25$
   b. $2|2x - 3| + 1 = -|2x - 3| + 22$

9. Solve the following equations:
a. $|3x^2 - 7x - 6| = 0$
b. $|x^2 + 14x - 68| = 4$

10. Solve the following equations:
a. $|5x^2 - 13x + 6| = 0$
b. $|2x^2 + 9x - 12| = 3$

CHAPTER 5
INEQUALITIES INVOLVING ABSOLUTE VALUES, QUOTIENT AND SQUARE FUNCTIONS

The modulus or absolute value of a number is the size of the number without its sign. It is denoted by a vertical line enclosing a number. For example |5| = 5 and |−5| = 5. A negative number enclosed in the vertical lines is regarded as positive.

If an inequality is given by : |x| < 2, it means that x is in the range: −2 < x < 2, i.e. x < 2 and x > −2

Examples

1. Solve the inequality: |5x − 2| < 13

<u>Solution</u>

|5x − 2| < 13

Note that this also means |−(5x −2)| < 13.

The range of the term in modulus is given by:

−13 < 5x − 2 < 13

We now take them separately and solve. Taking the first part (left hand part) gives:

−13 < 5x − 2

−13 + 2 < 5x

−11 < 5x

$-\dfrac{11}{5} < x$

Or $x > -\dfrac{11}{5}$ (Make sure the open side of the inequality sign still faces x when rearranging it).

Let us now take the other part of the inequality range and solve. This gives:

5x − 2 < 13

5x < 13 + 2

5x < 15

$x < \dfrac{15}{5}$

$x < 3$

We now combine the two results to obtain the range of the solution as follows:

$-\dfrac{11}{5} < x < 3$

Note that the opened end of the inequality sign is facing x in the solution $x > -\dfrac{11}{5}$, while the closed end (elbow end) of the inequality sign is facing x in the solution $x < 3$. Therefore, when combining the solution (i.e. $-\dfrac{11}{5} < x < 3$) you have to ensure that the right part of the inequality sign is facing x.

2. Solve the inequality $|3x + 7| \geq 10$

Solution

$|3x + 7| \geq 10$

This means: $(3x + 7) \geq 10$. This directly gives: $3x + 7 \geq 10$

It also means $-(3x + 7) \geq 10$. When we divide both sides by -1, it gives

$$\frac{-(3x + 7)}{-1} \geq \frac{10}{-1}$$

∴ $3x + 7 \leq -10$

Take note of the reversal of the inequality sign since both sides of the inequality were divided by a negative number.

Hence, the two inequalities obtained from $|3x + 7| \geq 10$ are:

$3x + 7 \geq 10$ and $3x + 7 \leq -10$

We now take each inequality and solve. This gives:

$3x + 7 \geq 10$

$3x \geq 10 - 7$

$3x \geq 3$

∴ $x \geq \frac{3}{3}$

$x \geq 1$

Solving the second inequality obtained gives:

$3x + 7 \leq -10$

$3x \leq -10 - 7$

$3x \leq -17$

$x \leq -\frac{17}{3}$

$x \leq -5\frac{2}{3}$

∴ $|3x + 7| \geq 10$ is satisfied when $x \geq 1$ and $x \leq -5\frac{2}{3}$

Note that these two results cannot be combined together since the solutions do not range from one to the other. This is because $x \geq 1$ means x can be 1, 2, 3 ..., while $x \leq -5\frac{2}{3}$ means that x can be $-5\frac{2}{3}$, -6, -7.... Hence, the two set of values do not meet and so cannot be combined together. A solution such as $x \leq 1$ and $x \geq -3$ can be combined together to give $-3 \leq x \leq 1$. This is because the values meet as follows: $-3, -2, -1, 0, 1$. Therefore, take note of this point before combining any pair of solutions.

3. Solve: $|4x - 9| \leq 3$

Solution

$|4x - 9| \leq 3$

The two inequalities that can be obtained from this are:

$4x - 9 \leq 3$

And $4x - 9 \geq -3$ (This is obtained by simply reversing the inequality sign and changing the positive number on the right to be negative)

We now solve each of the inequality as follows:

$4x - 9 \leq 3$

$4x \leq 3 + 9$

$4x \leq 12$

$x \leq \dfrac{12}{4}$

$x \leq 3$

And: $4x - 9 \geq -3$

$4x \geq -3 + 9$

$4x \geq \dfrac{6}{4}$

$x \geq \dfrac{3}{2}$

Wee can now combine the solutions to obtain: $\dfrac{3}{2} \leq x \leq 3$

$\therefore$ The inequality $|4x - 9| \leq 3$ is satisfied within the range $\dfrac{3}{2} \leq x \leq 3$

4. Solve the inequality $|3m - 5| > 4$

Solution

$|3m - 5| > 4$

The two inequalities that can be obtained from this are:

$3m - 5 > 4$

And $3m - 5 < -4$ (Simply reverse the inequality sign and assign a negative sign to the number on the right hand side)

We now solve each of the inequality as follows:

$3m - 5 > 4$

$3m > 4 + 5$

$3m > 9$

$m > \dfrac{9}{3}$

$m > 3$

And $3m - 5 < -4$

$3m < -4 + 5$

$3m < 1$

$m < \dfrac{1}{3}$

$\therefore$ The inequality $|3m - 5| > 4$ is satisfied when $m > 3$ and $m < \dfrac{1}{3}$ (They cannot be combined)

Inequalities Involving Quotients

If $\frac{x}{y} > 0$, then $\frac{x}{y}$ has to be a positive value.

For $\frac{x}{y}$ to be positive either both x and y must be positive or both x and y must be negative.

If $\frac{x}{y} < 0$, then $\frac{x}{y}$ has to be a negative value. For $\frac{x}{y}$ to be negative, only x or only y must be negative.

However, for inequalities such as $\frac{x}{y} \geq 0$ or $\frac{x}{y} \leq 0$, the same rules above applies to the numerator (i.e. x), but $y \geq 0$ or $y \leq 0$ cannot be used for the denominator so that it will not render the fraction undefined. Hence we use $y > 0$ or $y < 0$ for the denominator. This is because a fraction such as $\frac{2}{0}$ is undefined. So, in $\frac{x}{y} \geq 0$ or $\frac{x}{y} \leq 0$, $y \geq 0$ or $y \leq 0$ is not possible since it will make the fraction undefined.

Examples

1. Solve the inequality $\frac{x-2}{3x+9} > 0$

Solution

$$\frac{x-2}{3x+9} > 0$$

For the inequality above to be greater than zero, (i.e. > 0) it means that it has to be positive. For it to be positive, either the numerator and denominator are both positive (i.e. > 0) or they are both negative (i.e. < 0). This means that:

a. When both numerator and denominator are positive, then:

$x - 2 > 0$

and $3x + 9 > 0$

Solving each of these gives:

$x - 2 > 0$

$\therefore \quad x > 2$

and $3x + 9 > 0$

$\quad 3x > -9$

$\quad x > -\frac{9}{3}$

$\quad x > -3$

Therefore, we now have two possible solutions. But we have to test each of the solution to see if they are true. This is done by substituting a possible value in each solution into the original inequality. Let us test the solution $x > 2$. A possible value here is 3 (since $3 > 2$). Now substitute 3 for x in the original inequality. This gives:

$$\frac{x-2}{3x+9} > 0$$

$$\frac{3-2}{3(3)+9} > 0$$
$$\frac{1}{9+9} > 0$$
$$\frac{1}{18} > 0 \quad \text{(This is true)}$$

Hence, the solution $x > 2$ is correct.

Let us test the solution $x > -3$. A possible value here is -2 (Since $-2 > -3$). Now substitute -2 for x in the original inequality. This gives:

$$\frac{x-2}{3x+9} > 0$$
$$\frac{-2-2}{3(-2)+9} > 0$$
$$\frac{-4}{-6+9} > 0$$
$$\frac{-4}{3} > 0$$
$$-\frac{4}{3} > 0 \quad \text{(This is not true since 0 is greater)}$$

Hence, the solution $x > -3$ is not correct.

b. Now we are through with when numerator and denominator are positive. When numerator and denominator are negative, then:

$$x - 2 < 0$$
And $\quad 3x + 9 < 0$

Solving each of them gives:

$$x - 2 < 0$$
$$x < 2$$

And $\quad 3x + 9 < 0$
$$3x < -9$$
$$x < -\frac{9}{3}$$
$$x < -3$$

Hence, $x < 2$ or $x < -3$

Let us now test each of the solution to see if they are true.

When $x < 2$, a possible value here is 1 (since $1 < 2$). Now substitute 1 for x in the inequality. This gives:

$$\frac{x-2}{3x+9} > 0$$
$$\frac{1-2}{3(1)+9} > 0$$
$$\frac{-1}{3+9} > 0$$
$$-\frac{1}{12} > 0 \quad \text{(This is not true since 0 is greater than } -\frac{1}{12}\text{)}$$

46

Hence, the solution $x < 2$ is not correct.

Similarly, when $x < -3$, a possible value here is -4 (since $-4 < -3$). We now substitute -4 for x in the inequality. This gives:

$$\frac{x-2}{3x+9} > 0$$

$$\frac{-4-2}{3(-4)+9} > 0$$

$$\frac{-6}{-12+9} > 0$$

$$\frac{-6}{-3} > 0$$

$$2 > 0 \quad \text{(This is true)}$$

Hence, the solution $x < -3$ is correct.

Hence, in all our four solutions, the two that are correct are $x > 2$ and $x < -3$

Therefore, $\frac{x-2}{3x+9} > 0$ is satisfied when $x > 2$ and $x < -3$

2. Solve the inequality $\frac{2m+6}{m-4} < 0$

Solution

$$\frac{2m+6}{m-4} < 0$$

Example (1) above appears complex because I took my time to thoroughly explain it. In order to avoid making this example as long as example (1), I will use a more direct procedure.

Here, let us solve each of the numerators and denominators using the equality sign as follows:

$2m + 6 = 0$

$2m = -6$

$m = -\frac{6}{2}$

$m = -3$

Applying inequality to the solution above implies that:

$m < -3$ or $m > -3$ (Use the two inequality signs)

Let us test each of them to find the correct solution.

When $m < -3$, a possible value here is -3.5 (since $-3.5 < -3$. Testing with a value that just begins the solution puts us on a safer side). Now, substitute -3.5 into the inequality. This gives:

$$\frac{2m+6}{m-4} < 0$$

$$\frac{2(-3.5)+6}{-3.5-4} < 0$$

$$\frac{-7+6}{-3.5-4} < 0$$

$$\frac{-1}{-7.5} < 0$$
$$\frac{1}{7.5} < 0 \quad \text{(This is not true)}$$

When m > −3, a possible value here is −2.5. Substituting this value into the inequalities gives:
$$\frac{2m+6}{m-4} < 0$$
$$\frac{2(-2.5)+6}{-2.5-4} < 0$$
$$\frac{-5+6}{-2.5-4} < 0$$
$$\frac{1}{-6.5} < 0 \quad \text{(This is true)}$$

Hence m > −3 is correct.

Let us now carry out the same procedure with the denominator.

    m − 4 = 0

∴   m = 4

Hence, m < 4 or m > 4   (When the two inequality signs are applied)

When m < 4, let us substitute 3.5 into the inequality as follows:
$$\frac{2m+6}{m-4} < 0$$
$$\frac{2(3.5)+6}{3.5-4} < 0$$
$$\frac{7+6}{3.5-4} < 0$$
$$\frac{13}{-0.5} < 0$$
$$-26 < 0 \quad \text{(This is true)}$$

Hence, m < 4 is correct

Since this is correct, it means that the other solution of m > 4 is wrong since we already have our two possible solutions. However, let us test the other value.

When m > 4, let us substitute 4.5 into the inequality as follows:
$$\frac{2m+6}{m-4} < 0$$
$$\frac{2(4.5)+6}{4.5-4} < 0$$
$$\frac{9+6}{0.5} < 0$$
$$\frac{15}{0.5} < 0$$
$$30 < 0 \quad \text{(This is not true)}$$

Therefore, our two possible solutions are m > −3 or m < 4. They can be combined to give:

    −3 < x < 4

Therefore, $\frac{2m+6}{m-4} < 0$, is satisfied when −3 < x < 4

3. Solve the inequality $\dfrac{2x-3}{x+2} \leq 1$

Solution

$$\dfrac{2x-3}{x+2} \leq 1$$

Let us first make the right hand side to be zero.

$$\therefore \dfrac{2x-3}{x+2} - 1 \leq 0$$

Simplifying with $x+2$ as the LCM gives:

$$\dfrac{2x-3-1(x+2)}{x+2} \leq 0$$

$$\dfrac{2x-3-x-2}{x+2} \leq 0$$

$$\dfrac{2x-x-3-2}{x+2} \leq 0$$

$$\dfrac{x-5}{x+2} \leq 0$$

At this point, we can now take the numerator and solve for x as follows:

$$x - 5 = 0$$
$$x = 5$$

Applying inequality to this solution gives:

$$x \leq 5 \text{ or } x \geq 5$$

If we test the inequality when $x = 4.5$ (for $x \leq 5$) and $x = 5.5$ (for $x \geq 5$) respectively, we will find out that the true solution is $x \leq 5$

Taking the denominator gives:

$$x + 2 = 0$$
$$x = -2$$

Applying inequality to this solution gives:

$$x < -2 \text{ or } x > -2$$

Note that we cannot use $\leq$ or $\geq$ for the denominator since the denominator cannot be zero. If this is done, it will make the denominator undefined since fractions such as $\dfrac{5}{0}$ is undefined.

Now, back to $x < -2$ or $x > -2$. If we test the inequality with $x = -2.5$ (for $x < -2$) and $x = -1.5$ (for $x > -2$), respectively, we will find out that the true solution is $x > -2$.

$\therefore$ The solutions are $x \leq 5$ or $x > -2$

Hence, $\dfrac{2x-3}{x+2} \leq 1$ is satisfied when $-2 < x \leq 5$

4. Solve the inequality $\dfrac{2-b}{b+3} \geq 4$

Solution
$$\frac{2-b}{b+3} \geq 4$$
Let us first make the right hand side to be zero.
$$\frac{2-b}{b+3} \geq 4$$
$$\frac{2-b}{b+3} - 4 \geq 0$$
$$\frac{2-b-4(b+3)}{b+3} \geq 0$$
$$\frac{2-b-4b-12}{b+3} \geq 0$$
$$\frac{-5b-10}{b+3} \geq 0$$
Taking the numerator gives:
$$-5b - 10 = 0$$
$$-5b = 10$$
$$b = \frac{10}{-5}$$
$$b = -2$$
∴ $b \geq -2$ or $b \leq -2$

Testing possible values (−1.5 and −2.5) in each of these two solutions shows that $b \leq -2$ is the correct solution.

Taking the denominator gives:
$$b + 3 = 0$$
$$b = -3$$
∴ $b < -3$ or $b > -3$ (Note that $\geq$ or $\leq$ cannot be used for the denominator)

Testing values (−3.5 and −2.5) from each of the solution above shows that $b > -3$ is the correct solution.

Hence the solution are $b \leq -2$ or $b > -3$

Therefore, $\frac{2-b}{b+3} \geq 4$ is satisfied when $-3 < b \leq -2$

Inequalities Involving Square Functions

If $x^2 < c$, then $x < \sqrt{c}$ or $x > -\sqrt{c}$

Similarly, if $x^2 > c$, then $x > \sqrt{c}$ or $x < -\sqrt{c}$

These rules are applied in inequalities involving square functions.

Examples
1. Solve the inequality $x^2 > 16$

Solution

$$x^2 > 16$$

$\therefore$ $x > \sqrt{16}$

$x > 4$

Or $x < -\sqrt{16}$ (Reverse the inequality sign and introduce a negative sign)

$x < -4$

$\therefore$ $x > 4$ or $x < -4$

Therefore $x^2 > 16$ is satisfied when $x > 4$ or $x < -4$

2. Solve the inequality: $y^2 < 25$

Solution

$$y^2 < 25$$

$\therefore$ $y < \sqrt{25}$

$y < 5$

Or $y > -\sqrt{25}$ (Reverse the inequality sign and introduce a negative sign)

$y > -5$

Therefore $y^2 < 25$ is satisfied when $-5 < y < 5$

3. Solve: $(2a - 3)^2 \geq 49$

Solution

$$(2a - 3)^2 \geq 49$$

$2a - 3 \geq \sqrt{49}$

$2a - 3 \geq 7$

$2a \geq 7 + 3$

$2a \geq 10$

$a \geq \dfrac{10}{2}$

$a \geq 5$

Or $(2a - 3) \leq -\sqrt{49}$ (Reverse the inequality sign and introduce a negative sign)

$2a - 3 \leq -7$

$2a \leq -7 + 3$

$2a \leq -4$

$a \leq -\dfrac{4}{2}$

$a \leq -2$

Therefore $(2a - 3)^2 \geq 49$ is satisfied when $x \geq 5$ or $x \leq -2$

4. Solve: $7 - 4y^2 \leq -29$

Solution

$7 - 4y^2 \leq -29$

$-4y^2 \leq -29 - 7$

$-4y^2 \leq -36$

$y^2 \geq \dfrac{-36}{-4}$ (Take note of the reversal of the inequality sign due to division by a negative number)

$y^2 \geq 9$

Hence, $y \geq \sqrt{9}$

$y \geq 3$

Or $y \leq -\sqrt{9}$

$y \leq -3$

$\therefore$ $7 - 4y^2 \leq -29$ is satisfied when $y \geq 3$ or $y \leq -3$

5. Solve $5m^2 + 2 \geq 12$

Solution

$5m^2 + 2 \geq 12$

$5m^2 \geq 12 - 2$

$5m^2 \geq 10$

$m^2 \geq \dfrac{10}{5}$

$m^2 \geq 2$

$\therefore$ $m \geq \sqrt{2}$

Or $m \leq -\sqrt{2}$

Therefore $5m^2 + 2 \geq 12$ is satisfied when $m \geq \sqrt{2}$ or $m \leq -\sqrt{2}$

Exercise 5

Solve the following inequalities:

1. $|5x + 3| \geq 7$
2. $|3x - 2| < 17$
3. $|x - 8| \leq 11$
4. $|7m + 5| > 2$
5. $\dfrac{x - 5}{2x + 6} > 0$
6. $\dfrac{3m + 9}{m - 2} < 0$
7. $\dfrac{2x - 7}{x - 2} \leq 1$

8. $\dfrac{9-x}{2x+4} \geq 5$
9. $2x^2 > 18$
10. $y^2 < 36$
11. $(3a-5)^2 \geq 125$
12. $5 - 2y^2 \leq -13$
13. $9m^2 - 1 \geq 8$
14. $\dfrac{19-2x}{2x+1} \geq 3$
15. $7 - 5y^2 \leq -73$

CHAPTER 6
INDICIAL EQUATIONS

An indicial equation is an equation in which the unknown variable is a power (index) in the equation. Usually the knowledge of indices and logarithms is applied in solving this type of equation.

Examples
1. Solve: $3^{3x+4} = 27^{2x+5}$

<u>Solution</u>
$$3^{3x+4} = 27^{2x+5}$$

In order to solve indicial equation, express both sides of the equation in the same base, and then equate the powers. The left hand side of the equation above has a base of 3. It is clear that the right hand side can also be expressed as a base of 3 because $27 = 3^3$. Hence, the equation now simplifies as follows:

$$3^{3x+4} = (3^3)^{2x-5}$$
$$3^{3x+4} = 3^{3(2x-5)} \quad \text{(From indices, } (a^x)^y = a^{xy})$$
$$3^{3x+4} = 3^{6x-15}$$

Since the bases on both sides of the equation are equal, it implies that the powers are also equal. Hence we ignore the bases and equate their powers as follows:

$$3x + 4 = 6x - 15$$
$$4 + 15 = 6x - 3x$$
$$19 = 3x$$
$$\therefore \quad x = \frac{19}{3}$$
$$x = 6\frac{1}{3}$$

2. Solve the equation: $\frac{8^{x+3}}{4^{3x-1}} = 32^{2x+7}$

<u>Solution</u>
$$\frac{8^{x+3}}{4^{3x-1}} = 32^{2x+7}$$

We can express each of the bases in the equation as a base of two. This gives:

$$\frac{(2^3)^{x+3}}{(2^2)^{3x-1}} = (2^5)^{2x+7} \quad \text{(Note that } 8 = 2^3, 4 = 2^2 \text{ and } 32 = 2^5)$$

$$\frac{2^{3(x+3)}}{2^{2(3x-1)}} = 2^{5(2x+7)} \quad \text{(Since } (a^x)^y = a^{xy})$$

$$\frac{2^{3x+9}}{2^{6x-2}} = 2^{10x+35}$$

∴ $2^{3x+9-(6x-2)} = 2^{10x+35}$ (From indices, $\frac{a^x}{a^y} = a^{x-y}$, i.e. subtraction of powers)

$2^{3x+9-6x+2} = 2^{10x+35}$

$2^{-3x+11} = 2^{10x+35}$

Since the bases are equal, it also means that the powers are equal. Hence, we equate the powers as follows:

$-3x + 11 = 10x + 35$

$-3x - 10x = 35 - 11$

$-13x = 24$

∴ $x = -\frac{24}{13}$

$x = -1\frac{11}{13}$

3. Solve: $\frac{125^{x-5}}{625^{2x}} = \frac{5^{3x+1}}{25^{6x-1}}$

Solution

$\frac{125^{x-5}}{625^{2x}} = \frac{5^{3x+1}}{25^{6x-1}}$

Let us express each base above as a base of 5 as follows:

$\frac{(5^3)^{x-5}}{(5^4)^{2x}} = \frac{5^{3x+1}}{(5^2)^{6x-1}}$

$\frac{5^{3(x-5)}}{5^{4(2x)}} = \frac{5^{3x+1}}{5^{2(6x-1)}}$

$\frac{5^{3x-15}}{5^{8x}} = \frac{5^{3x+1}}{5^{12x-2}}$

Cross multiply to obtain:

$5^{3x-15} \times 5^{12x-2} = 5^{3x+1} \times 5^{8x}$

$5^{3x-15+12x-2} = 5^{3x+1+8x}$ (From indices, $a^x \times a^y = a^{x+y}$ i.e. addition of powers)

$5^{15x-17} = 5^{11x+1}$

We now equate the powers since the bases are equal.

∴ $15x - 17 = 11x + 1$

$15x - 11x = 1 + 17$

$4x = 18$

$x = \frac{18}{4}$

$= \frac{9}{2}$

∴ $x = 4\frac{1}{2}$

4. Solve the equation $3^{2x} = 12$

Solution

$$3^{2x} = 12$$

A careful look at this equation shows that the two sides of the equation cannot be expressed in the same base. Hence, we take the logarithm of both sides of the equation as follows:

$3^{2x} = 12$

$\log 3^{2x} = \log 12$

$\therefore$ $2x \log 3 = \log 12$ (Note that from logarithm, $\log x^y = y \log x$)

$2x (0.4771) = 1.0792$ (From calculator, $\log 3 = 0.4771$ and $\log 12 = 1.0792$)

$0.9542x = 1.0792$

$\therefore$ $x = \dfrac{1.0792}{0.9542}$

$x = 1.13$

5. Solve the equation $5^{4x-1} = 18^{2x+3}$

Solution

$$5^{4x-1} = 18^{2x+3}$$

It is clear that 5 and 18 cannot be expressed in the same base in whole number. Hence, we take the logarithm of both sides of the equation as follows:

$5^{4x-1} = 18^{2x+3}$

$\log 5^{4x-1} = \log 18^{2x+3}$

$(4x - 1)\log 5 = (2x + 3)\log 18$

$(4x - 1)(0.6990) = (2x + 3)(1.2553)$ (Note that $\log 5 = 0.6990$ and $\log 18 = 1.2553$)

$2.796x - 0.6990 = 2.5106x + 3.7659$

$2.796x - 2.5106x = 3.7659 + 0.669$

$0.2854x = 4.4649$

$\therefore$ $x = \dfrac{4.4649}{0.2854}$

$x = 15.64$

6. Solve $3.2^x \times 7.53^{3x+2} = 4^{2x+5}$

Solution

$3.2^x \times 7.53^{3x+2} = 4^{2x+5}$

Taking the logarithm of both sides gives:

$\log(3.2^x \times 7.53^{3x+2}) = \log 4^{2x+5}$

$\log 3.2^x + \log 7.53^{3x+2} = \log 4^{2x+5}$ (Note that from logarithm, $\log(AB) = \log A + \log B$)

$x \log 3.2 + (3x + 2)\log 7.53 = (2x + 5)\log 4$

$x(0.5051) + (3x + 2)(0.8573) = (2x + 5)(0.6021)$

$$0.5051x + 2.5719x + 1.7146 = 1.2042x + 3.0105$$
$$0.5051x + 2.5719x - 1.2042x = 3.0105 - 1.7146$$
$$1.8728x = 1.2959$$
$$x = \frac{1.2959}{1.8728}$$
$$x = 0.692$$

7. Solve the equation $2^{x^2} = 4^{3x-4}$

Solution
$$2^{x^2} = 4^{3x-4}$$
Expressing both sides in the same base (i.e. 2) gives:
$$2^{x^2} = (2^2)^{3x-4}$$
$$2^{x^2} = 2^{2(3x-4)}$$
$$2^{x^2} = 2^{6x-8}$$
Since the bases are equal, we now equate the powers as follows:
$$x^2 = 6x - 8$$
$$x^2 - 6x + 8 = 0$$
Solving this equation by factorization gives:
$$(x-4)(x-2) = 0$$
This gives: $x = 4$, or $x = 2$

8. Solve the equation $9^{x^2} = 3^{5x-3}$

Solution
$$9^{x^2} = 3^{5x-3}$$
Expressing both sides in the same base of 3 gives:
$$(3^2)^{x^2} = 3^{5x-3}$$
$$3^{2x^2} = 3^{5x-3}$$
Equating the powers gives:
$$2x^2 = 5x - 3$$
$$2x^2 - 5x + 3 = 0$$
Solving this equation by factorization gives:
$$2x^2 - 2x - 3x + 3 = 0$$
$$2x(x-1) - 3(x-1) = 0$$
$$(x-1)(2x-3) = 0$$
$$\therefore \quad x = 1 \text{ or } x = \frac{3}{2}$$

9. Solve the equation: $4^x - 3 \times 2^x + 2 = 0$

Solution

This equation is slightly different from others since it contains the addition/subtraction of three terms. The terms having x as their powers can be expressed in the same base as follows.

$(2^2)^x - 3(2^x) + 2 = 0$

$(2^x)^2 - 3(2^x) + 2 = 0$ (Note that $(2^2)^x = (2^x)^2$, since 2 x x is the same as x x 2)

Since $(2^x)^2$ means (2^x) raise to the power 2, it means that this equation can be expressed as a quadratic equation.

Let $2^x = y$ ....Equation (1)

Substitute y for 2^x in the simplified equation above as follows

$(2^x)^2 - 3(2^x) + 2 = 0$

$y^2 - 3y + 2 = 0$ (Since $2^x = y$)

Solving this equation by factorization gives:

$(y - 1)(y - 2) = 0$

Hence, y = 1 or 2

When y = 1, we substitute 1 for y in equation (1) in order to find x. This gives:

$2^x = y$ ....Equation (1)

$2^x = 1$

$2^x = 2^0$ (Note that $2^0 = 1$)

Equating the powers gives:

$x = 0$

When y = 2, we substitute 2 for y in equation (1) in order to find x. This gives:

$2^x = y$ ....Equation (1)

$2^x = 2$

$2^x = 2^1$ (Note that $2^1 = 2$)

Equating the powers gives:

$x = 1$

Hence, $x = 0$ or $x = 1$.

10. Solve the equation $2^{2x} - 2^{1+x} - 8 = 0$

Solution

$2^{2x} - 2^{1+x} - 8 = 0$

In this equation, 2^{1+x} can be expressed as 2^1 x 2^x, since from indices a^x x $a^y = a^{x+y}$. This also means that $a^{x+y} = a^x$ x a^y. Hence substituting 2^1 x 2^x for 2^{1+x} gives:

$2^{2x} - 2^{1+x} - 8 = 0$

$2^{2x} - 2^1$ x $2^x - 8 = 0$

$(2^x)^2 - 2(2^x) - 8 = 0$

Now, let $2^x = b$ ....Equation (1)

Substitute b for 2^x in the equation above. This gives:

$(2^x)^2 - 2(2^x) - 8 = 0$
$b^2 - 2b - 8 = 0$ (Since $2^x = b$)

Solving this equation by factorization gives:
$(b - 4)(b + 2) = 0$
∴ $b = 4$ or $b = -2$

When $b = 4$, we substitute 4 for b in equation (1). This gives:
$2^x = b$ ....Equation (1)
$2^x = 4$
∴ $2^x = 2^2$

Equating powers gives:
$x = 2$

Note that the other solution of $b = -2$ has been discarded since substituting -2 for b in equation (1) cannot be solved, i.e.:
$2^x = -2$ has no solution

11. Solve: $3^{1+2x} + 2 \times 3^x - 1 = 0$

Solution
$3^{1+2x} + 2 \times 3^x - 1 = 0$
$3^1 \times 3^{2x} + 2 \times 3^x - 1 = 0$ (Note that $3^1 \times 3^{2x} = 3^{1+2x}$ from indices)
$3(3^x)^2 + 2(3^x) - 1 = 0$

Let $3^x = y$ ....Equation (1)

Substitute y for 3^x in the equation above. This gives:
$3(3^x)^2 + 2(3^x) - 1 = 0$
$3y^2 + 2y - 1 = 0$ (Since $3^x = y$)

Solving this equation by factorization gives:
$3y^2 + 3y - y - 1 = 0$
$3y(y + 1) - 1(y + 1) = 0$
$(y + 1)(3y - 1) = 0$
∴ $y = -1$ or $y = \frac{1}{3}$

When $y = \frac{1}{3}$, we substitute $\frac{1}{3}$ for y in equation (1). This gives:
$3^x = y$ ....Equation (1)
$3^x = \frac{1}{3}$
∴ $3^x = 3^{-1}$ (Note that $3^{-1} = \frac{1}{3}$)

Equating powers gives:
$x = -1$

Note that the other solution of $y = -1$ has been discarded since we cannot use it to solve for x.

12. Solve the simultaneous equation: $9^{2x-y} = 3$ and $16^{x+y} = 8$

Solution

Let us simplify each of the equation above one after the other.
$$9^{2x-y} = 3$$
Expressing both sides of the equation in the same base gives:
$$(3^2)^{2x-y} = 3^1$$
$$3^{2(2x-y)} = 3^1$$
$$3^{4x-2y} = 3^1$$
Equating the powers gives:
$$4x - 2y = 1 \quad \text{.....................Equation (1)}$$
Similarly, $16^{x+y} = 8$ can be expressed in base 2 as follows:
$$(2^4)^{x+y} = 2^3$$
$$2^{4(x+y)} = 2^3$$
$$2^{4x+4y} = 3^3$$
Equating the powers gives:
$$4x + 4y = 3 \quad \text{.....................Equation (2)}$$
Bring equation 1 and 2 together in order to solve them simultaneously. This gives:
$$4x - 2y = 1 \quad \text{.....................Equation (1)}$$
$$4x + 4y = 3 \quad \text{.....................Equation (2)}$$
Equation (2) – (1): $6y = 2$ (Note that $4y - (-2y) = 6y$, and $3 - 1 = 2$)
$$y = \frac{2}{6}$$
$$y = \frac{1}{3}$$
Substitute $\frac{1}{3}$ for y in equation (1).
$$4x - 2y = 1 \quad \text{.....................Equation (1)}$$
$$4x - 2(\frac{1}{3}) = 1$$
$$4x - \frac{2}{3} = 1$$
$$4x = 1 + \frac{2}{3}$$
$$4x = 1\frac{2}{3}$$
$$4x = \frac{5}{3}$$
$$x = \frac{\frac{5}{3}}{4}$$
$$= \frac{5}{3} \div 4$$

$$= \frac{5}{3} \times \frac{1}{4} \quad \text{(Note that 4 also means } \frac{4}{1}\text{)}$$
$$x = \frac{5}{12}$$
$$\therefore \quad x = \frac{5}{12} \text{ and } y = \frac{1}{3}$$

13. Solve simultaneously, the equations: $4^{x-2y} = 64$ and $25^{4x-3y} = 625$

Solution
$$4^{x-2y} = 64$$
Expressing both sides of the equation in the same base gives:
$$4^{x-2y} = 4^3$$
Equating the powers gives:
$\quad x - 2y = 3$ ....Equation (1)

Similarly, $25^{4x-3y} = 625$ can be expressed in base 25 as follows:
$$25^{4x-3y} = 25^2$$
Equating the powers gives:
$\quad 4x + 3y = 2$ ....Equation (2)

From equation (1), $x = 3 + 2y$ ....Equation (3)

Substitute $3 + 2y$ for x in equation (2). This gives:
$\quad 4x + 3y = 2$ ....Equation (2)
$\quad 4(3 + 2y) + 3y = 2$
$\quad 12 + 8y + 3y = 2$
$\quad 11y = 2 - 12$
$\quad 11y = -10$
$\quad y = \frac{-10}{11}$
$\quad y = -\frac{10}{11}$

Substitute $-\frac{10}{11}$ for y in equation (3) (Note that any of the equations can be used)
$\quad x = 3 + 2y$ ....Equation (3)
$\quad\quad = 3 + 2(-\frac{10}{11})$
$\quad\quad = 3 - \frac{20}{11}$
$\quad x = \frac{33 - 20}{11}$

$$x = \frac{13}{11}$$
$$x = 1\frac{2}{11}$$
$$\therefore \quad x = 1\frac{2}{11} \text{ and } y = -\frac{10}{11}$$

Exercise 6

1. Solve: $2^{5x+1} = 32^{2x-3}$
2. Solve the equation: $\dfrac{4^{2x+5}}{16^{2x-7}} = 64^{3x+2}$
3. Solve: $\dfrac{6^{3x-5}}{216^x} = \dfrac{216^{1-3x}}{36^{2x-3}}$
4. Solve the equation $5^{4x} = 15$
5. Solve the equation $3^{2x-11} = 20^{x+7}$
6. Solve $4.5^{2x} \times 10.2^{2-5x} = 8^{9x-4}$
7. Solve the equation $5^{x^2} = 2^{5x+2}$
8. Solve the equation $25^{x^2} = 125^{3x-1}$
9. Solve the equation: $4^x - 3 \times 2^x - 40 = 0$
10. Solve the equation $3^{2x} - 3^{x-1} - 78 = 0$
11. Solve: $5^{1+2x} + 2 \times 3^x - 7 = 0$
12. Solve the simultaneous equation: $9^{x-y} = 81$ and $25^{2x+y} = 25$
13. Solve simultaneously, the equations: $2^{5x-3y} = 32$ and $27^{2x-y} = 9^x$
14. Solve the simultaneous equation: $4^{x-2y} = 64$ and $25^{2x+5y} = 625$
15. Solve the equation $64^{x^2-2} = 16^{x+1}$

CHAPTER 7
ROOTS OF QUADRATIC EQUATIONS (USE OF ALPHA AND BETA)

A quadratic equation can have three cases of roots (solution of a quadratic equation). Consider a quadratic equation given by:

$$ax^2 + bx + c = 0$$

1. The roots are real and different if $b^2 - 4ac > 0$
2. The roots are real and equal if $b^2 - 4ac = 0$ or $b^2 = 4ac$
3. The roots are complex if $b^2 - 4ac < 0$

If each term of the quadratic equation above is divided by 'a', it gives:

$$x^2 + \frac{b}{a}x + \frac{c}{a} = 0 \quad \text{............... (1)}$$

If α and β are the roots of the equation, then the equation can be written as:

$$x^2 - (\alpha + \beta)x + \alpha\beta = 0 \quad \text{............... (2)}$$

Or

$$x^2 - (\text{sum of roots})x + (\text{product of roots}) = 0 \quad \text{............... (3)}$$

Comparing equations (1) to (3) above shows that:

Sum of roots = $\alpha + \beta = -\frac{b}{a}$

And product of roots = $\alpha\beta = \frac{c}{a}$

The roots of quadratic equations can be expressed as functions of α and β.
In order to apply α and β in quadratic equations, it is important to know the following identities.

1. $\alpha^2 + \beta^2 = (\alpha + \beta)^2 - 2\alpha\beta$
2. $(\alpha - \beta)^2 = (\alpha + \beta)^2 - 4\alpha\beta$
3. $\alpha - \beta = \sqrt{(\alpha + \beta)^2 - 4\alpha\beta}$
4. $\alpha^2 - \beta^2 = (\alpha + \beta)(\alpha - \beta)$
   $= (\alpha + \beta)\sqrt{(\alpha + \beta)^2 - 4\alpha\beta}$
5. $\alpha^3 - \beta^3 = (\alpha - \beta)^3 + 3\alpha\beta(\alpha - \beta)$
6. $\alpha^3 + \beta^3 = (\alpha + \beta)^3 - 3\alpha\beta(\alpha + \beta)$

Examples
1. If α and β are the roots of the quadratic equation $2x^2 - 7x + 3 = 0$, find:
a. $\alpha + \beta$
b. $\alpha\beta$
c. $\alpha^2 + \beta^2$
d. $\frac{\alpha}{\beta} + \frac{\beta}{\alpha}$

e. $\dfrac{1}{\alpha} + \dfrac{1}{\beta}$

f. $\dfrac{1}{\alpha^2} + \dfrac{1}{\beta^2}$

g. $\alpha^3 + \beta^3$

Solution

a. $2x^2 - 7x + 3 = 0$

Recall the form: $ax^2 + bx + c = 0$

Comparing the two equations above shows that:

$a = 2, b = -7, c = 3$

But $\alpha + \beta = -\dfrac{b}{a} = -\left(\dfrac{-7}{2}\right)$ (Take note of the use of negative sign)

$\therefore \ \alpha + \beta = \dfrac{7}{2}$

And $\alpha\beta = \dfrac{c}{a} = \dfrac{3}{2}$

b. $\alpha\beta = \dfrac{3}{2}$ as shown in (a) above

c. $\alpha^2 + \beta^2$

Recall that $\alpha^2 + \beta^2 = (\alpha + \beta)^2 - 2\alpha\beta$, as given in the identities above.

Hence, $\alpha^2 + \beta^2 = (\alpha + \beta)^2 - 2\alpha\beta$

$\qquad = \left(\dfrac{7}{2}\right)^2 - 2\left(\dfrac{3}{2}\right)$ (By substituting $\dfrac{7}{2}$ for $(\alpha + \beta)$ and $\dfrac{3}{2}$ for $\alpha\beta$)

$\qquad = \dfrac{49}{4} - 3$

$\qquad = \dfrac{49 - 12}{4}$

$\qquad = \dfrac{37}{4}$

d. $\dfrac{\alpha}{\beta} + \dfrac{\beta}{\alpha} = \dfrac{\alpha^2 + \beta^2}{\alpha\beta}$ (Note that the LCM of α and β is $\alpha\beta$)

$\qquad = \dfrac{(\alpha + \beta)^2 - 2\alpha\beta}{\alpha\beta}$ (Note that $\alpha^2 + \beta^2 = (\alpha + \beta)^2 - 2\alpha\beta$)

$\qquad = \dfrac{\frac{37}{4}}{\frac{3}{2}}$ (Note that from (c) above $\alpha^2 + \beta^2 = \dfrac{37}{4}$)

$\qquad = \dfrac{37}{4} \times \dfrac{2}{3}$

$\qquad = \dfrac{37}{2} \times \dfrac{1}{3}$ (After division by 2)

$$= \frac{37}{6}$$

e. $\frac{1}{\alpha} + \frac{1}{\beta} = \frac{\beta + \alpha}{\alpha\beta}$

$$= \frac{\alpha + \beta}{\alpha\beta}$$

$$= \frac{\frac{7}{2}}{\frac{3}{2}}$$

$$= \frac{7}{2} \times \frac{2}{3}$$

$$= \frac{7}{3} \quad \text{(After 2 cancels out)}$$

f. $\frac{1}{\alpha^2} + \frac{1}{\beta^2} = \frac{\beta^2 + \alpha^2}{\alpha^2\beta^2}$

$$= \frac{\alpha^2 + \beta^2}{(\alpha\beta)^2}$$

$$= \frac{\frac{37}{4}}{(\frac{3}{2})^2}$$

$$= \frac{\frac{37}{4}}{\frac{9}{4}}$$

$$= \frac{37}{4} \times \frac{4}{9}$$

$$= \frac{37}{9} \quad \text{(Since 4 cancels out)}$$

g. $\alpha^3 + \beta^3 = (\alpha + \beta)^3 - 3\alpha\beta(\alpha + \beta)$

$$= (\frac{7}{2})^3 - 3(\frac{3}{2})(\frac{7}{2})$$

$$= \frac{343}{8} - \frac{9}{2}(\frac{7}{2})$$

$$= \frac{343}{8} - \frac{63}{4}$$

$$= \frac{343 - 126}{8}$$

$$= \frac{217}{8}$$

2. If the roots of the quadratic equation, $3x^2 + 5x - 9 = 0$, are α and β, find the equation whose roots are α^2 and β^2

Solution

$$3x^2 + 5x - 9 = 0$$

From this equation: a = 3, b = 5 and c = –9

But, $\alpha + \beta = -\dfrac{b}{a}$

$= -\dfrac{5}{3}$ (Since a = 3 and b = 5)

Also, $\alpha\beta = \dfrac{c}{a}$

$= \dfrac{-9}{3}$ (Since a = 3 and c = –9)

$= -3$

Now, roots of new equation are α^2 and β^2

∴ Sum of roots of new equation = $\alpha^2 + \beta^2$

$= (\alpha + \beta)^2 - 2\alpha\beta$ (From the identities given above)

$= \left(-\dfrac{5}{3}\right)^2 - 2(-3)$

$= \dfrac{25}{9} + 6$

$= \dfrac{25 + 54}{9}$

Sum of roots $(\alpha + \beta) = \dfrac{79}{9}$

And product of roots of new equation = $\alpha^2 \times \beta^2$

$= \alpha^2\beta^2$

$= (\alpha\beta)^2$

$= (-3)^2$

Product of roots $(\alpha\beta) = 9$

Recall that a quadratic equation is represented as:

$x^2 - (\text{sum of roots})x + (\text{product of roots}) = 0$

Or, $x^2 - (\alpha + \beta)x + (\alpha\beta) = 0$

When the values obtained for $\alpha + \beta$ and $\alpha\beta$ are substituted into the equation above, it gives:

$$x^2 - \dfrac{79}{9}x + 9 = 0$$

When each term is multiplied by 9 in order to clear out the fraction, it gives:

$$9x^2 - 79x + 81 = 0$$

∴ The equation whose roots are α^2 and β^2 is $9x^2 - 79x + 81 = 0$

3. If the roots of the quadratic equation $5x^2 + x + 4 = 0$, are α and β, find the equation whose roots are $\dfrac{\alpha}{\beta}$ and $\dfrac{\beta}{\alpha}$

Solution
$$5x^2 + x + 4 = 0$$
From this equation: a = 5, b = 1 and c = 4

But, $\alpha + \beta = -\dfrac{b}{a}$

$\qquad = -\dfrac{1}{5}$ (Since a = 5 and b = 1)

Also, $\alpha\beta = \dfrac{c}{a}$

$\qquad = \dfrac{4}{5}$ (Since a = 5 and c = 4)

Now, roots of new equation are $\dfrac{\alpha}{\beta}$ and $\dfrac{\beta}{\alpha}$

∴ Sum of roots of new equation $= \dfrac{\alpha}{\beta} + \dfrac{\beta}{\alpha}$

$= \dfrac{\alpha^2 + \beta^2}{\alpha\beta}$ (From the identities given above)

$= \dfrac{(\alpha+\beta)^2 - 2\alpha\beta}{\alpha\beta}$

$= \dfrac{(-\frac{1}{5})^2 - 2(\frac{4}{5})}{\frac{4}{5}}$

$= \dfrac{\frac{1}{25} - \frac{8}{5}}{\frac{4}{5}}$

$= \dfrac{\frac{1-40}{25}}{\frac{4}{5}}$

$= \dfrac{-\frac{39}{25}}{\frac{4}{5}}$

$= -\dfrac{39}{25} \times \dfrac{5}{4}$

$= -\dfrac{39}{5} \times \dfrac{1}{4}$ (After division by 5)

$= -\dfrac{39}{20}$

Sum of roots $(\alpha + \beta) = -\dfrac{39}{20}$

And product of roots of new equation $= \dfrac{\alpha}{\beta} \times \dfrac{\beta}{\alpha}$

$\qquad = \dfrac{\alpha\beta}{\alpha\beta}$

$\qquad = 1$

Product of roots $(\alpha\beta) = 1$

Recall that a quadratic equation is represented as:

$x^2 - $ (sum of roots)$x + $ (product of roots) $= 0$

Or, $x^2 - (\alpha + \beta)x + (\alpha\beta) = 0$

When the values obtained for $\alpha + \beta$ and $\alpha\beta$ are substituted into the equation above, it gives:

$$x^2 - (-\frac{39}{20})x + 1 = 0$$

$$x^2 + \frac{39}{20}x + 1 = 0$$

When each term is multiplied by 20 in order to clear out the fraction, it gives:

$$20x^2 + 39x + 20 = 0$$

∴ The equation whose roots are $\frac{\alpha}{\beta}$ and $\frac{\beta}{\alpha}$ is $20x^2 + 39x + 20 = 0$

4. Given that the quadratic equation $2x^2 - 5x - 3 = 0$, has roots α and β, find the equation whose roots are $\frac{1}{\alpha}$ and $\frac{1}{\beta}$

Solution

$$2x^2 - 5x - 3 = 0$$

From this equation: a = 2, b = −5 and c = −3

But, $\alpha + \beta = -\frac{b}{a}$

$= -(\frac{-5}{2})$ (Since a = 2 and b = −5)

$= \frac{5}{2}$

Also, $\alpha\beta = \frac{c}{a}$

$= \frac{-3}{2}$ (Since a = 2 and c = −3)

Now, roots of new equation are $\frac{1}{\alpha}$ and $\frac{1}{\beta}$

∴ Sum of roots of new equation $= \frac{1}{\alpha} + \frac{1}{\beta}$

$= \frac{\beta + \alpha}{\alpha\beta}$

$= \frac{\alpha + \beta}{\alpha\beta}$

$= \frac{\frac{5}{2}}{-\frac{3}{2}}$

$= \frac{5}{2} \times (-\frac{2}{3})$

$= -\frac{5}{3}$

Sum of roots of new equation $(\alpha + \beta) = -\frac{5}{3}$

And product of roots of new equation $= \frac{1}{\alpha} \times \frac{1}{\beta}$

$= \frac{1}{\alpha\beta}$

$= \frac{1}{-\frac{3}{2}}$

$= \frac{1}{1} \times (-\frac{2}{3})$

Product of roots of new equation $(\alpha\beta) = -\frac{2}{3}$

Recall that a quadratic equation is represented as:

$x^2 - (\text{sum of roots})x + (\text{product of roots}) = 0$

Or, $x^2 - (\alpha + \beta)x + (\alpha\beta) = 0$

When the values obtained for $\alpha + \beta$ and $\alpha\beta$ are substituted into the equation above, it gives:

$x^2 - (-\frac{5}{3})x + (-\frac{2}{3}) = 0$

$x^2 + \frac{5}{3}x - \frac{2}{3} = 0$

When each term is multiplied by 3 in order to clear out the fraction, it gives:

$3x^2 + 5x - 2 = 0$

∴ The equation whose roots are $\frac{1}{\alpha}$ and $\frac{1}{\beta}$ is $3x^2 + 5x - 2 = 0$

5. If α and β are the roots of the quadratic equation $x^2 - 7x + 9 = 0$, find the quadratic equation whose roots are $\frac{1}{\alpha^2}$ and $\frac{1}{\beta^2}$

Solution

$x^2 - 7x + 9 = 0$

From this equation: a = 1, b = –7 and c = 9

But, $\alpha + \beta = -\frac{b}{a}$

$= -(\frac{-7}{1})$ (Since a = 1 and b = –7)

$= 7$

Also, $\alpha\beta = \frac{c}{a}$

$= \frac{9}{1}$ (Since a = 1 and c = 9)

$= 9$

Now, roots of new equation are $\frac{1}{\alpha^2}$ and $\frac{1}{\beta^2}$

∴ Sum of roots of new equation = $\frac{1}{\alpha^2} + \frac{1}{\beta^2}$

$$= \frac{\alpha^2 + \beta^2}{\alpha^2 \beta^2}$$

$$= \frac{(\alpha+\beta)^2 - 2\alpha\beta}{(\alpha\beta)^2}$$

$$= \frac{7^2 - 2(9)}{9^2}$$

$$= \frac{49 - 18}{81}$$

Sum of roots $(\alpha + \beta) = \frac{31}{81}$

And product of roots of new equation = $\frac{1}{\alpha^2} \times \frac{1}{\beta^2}$

$$= \frac{1}{\alpha^2 \beta^2}$$

$$= \frac{1}{(\alpha\beta)^2}$$

$$= \frac{1}{9^2}$$

Product of roots $(\alpha\beta) = \frac{1}{81}$

Recall that a quadratic equation is represented as:

$x^2 -$ (sum of roots)$x +$ (product of roots) $= 0$

Or, $x^2 - (\alpha + \beta)x + (\alpha\beta) = 0$

When the values obtained for $\alpha + \beta$ and $\alpha\beta$ are substituted into the equation above, it gives:

$$x^2 - (\frac{31}{81})x + \frac{1}{81} = 0$$

When each term is multiplied by 81 in order to clear out the fractions, it gives:

$81x^2 - 31x + 1 = 0$

∴ The equation whose roots are $\frac{1}{\alpha^2} + \frac{1}{\beta^2}$ is $81x^2 - 31x + 1 = 0$

6. If α and β are the roots of the quadratic equation $2x^2 + 3x + 5 = 0$, find the quadratic equation whose roots are α^3 and β^3

Solution

$2x^2 + 3x + 5 = 0$

From this equation: a = 2, b = 3 and c = 5

But, $\alpha + \beta = -\dfrac{b}{a}$

$= -\dfrac{3}{2}$

Also, $\alpha\beta = \dfrac{c}{a}$

$= \dfrac{5}{2}$

Now, roots of new equation are α^3 and β^3

∴ Sum of roots of new equation = $\alpha^3 + \beta^3$

$= (\alpha + \beta)^3 - 3\alpha\beta(\alpha + \beta)$

$= (-\dfrac{3}{2})^3 - 3(\dfrac{5}{2})(-\dfrac{3}{2})$

$= -\dfrac{27}{8} + \dfrac{45}{4}$

$= \dfrac{-27 + 90}{8}$

$= \dfrac{63}{8}$

Sum of roots $(\alpha + \beta) = \dfrac{63}{8}$

And product of roots of new equation = $\alpha^3 \times \beta^3$

$= (\alpha\beta)^3$

$= (\dfrac{5}{2})^3$

$= \dfrac{125}{8}$

Product of roots of new equation $(\alpha\beta) = \dfrac{125}{8}$

Recall that a quadratic equation is represented as:

$x^2 - $ (sum of roots)$x + $ (product of roots) $= 0$

Or, $x^2 - (\alpha + \beta)x + (\alpha\beta) = 0$

When the values obtained for $\alpha + \beta$ and $\alpha\beta$ are substituted into the equation above, it gives:

$x^2 - (\dfrac{63}{8})x + \dfrac{125}{8} = 0$

When each term is multiplied by 8 in order to clear out the fractions, it gives:

$8x^2 - 63x + 125 = 0$

∴ The equation whose roots are α^3 and β^3 is $8x^2 - 63x + 125 = 0$

7. If α and β are the roots of the equation $4x^2 - 12x + 7 = 0$, find the values:
   a. $\alpha - \beta$
   b. $\alpha^2 - \beta^2$

Solution
$$4x^2 - 12x + 7 = 0$$
From this equation: a = 4, b = –12 and c = 7

But, $\alpha + \beta = -\dfrac{b}{a}$

$= -(\dfrac{-12}{4})$

$= 3$

Also, $\alpha\beta = \dfrac{c}{a}$

$= \dfrac{7}{4}$

Recall that: $(\alpha - \beta)^2 = (\alpha + \beta)^2 - 4\alpha\beta$

Taking the square root of both sides gives $\alpha - \beta$ as follows:

$\alpha - \beta = \sqrt{(\alpha + \beta)^2 - 4\alpha\beta}$

$= \sqrt{3^2 - 4(\dfrac{7}{4})}$

$= \sqrt{9 - 7}$

$= \sqrt{2}$

$\therefore \alpha - \beta = \sqrt{2}$ or 1.41

b. Recall that: $\alpha^2 - \beta^2 = (\alpha + \beta)(\alpha - \beta)$ (This is a difference of two squares)

$= (3)(\sqrt{2})$ (Since $\alpha + \beta = 3$ and $\alpha - \beta = \sqrt{2}$ from (a) above)

$= 3\sqrt{2}$

$\therefore \alpha^2 - \beta^2 = 3\sqrt{2}$ or 4.24

8. If α and β are the roots of the equation $3x^2 + 11x - 8 = 0$, find the value of $(\alpha - \beta)^2$

Solution
$$3x^2 + 11x - 8 = 0$$
From this equation: a = 3, b = 11 and c = –8

But, $\alpha + \beta = -\dfrac{b}{a}$

$= -\dfrac{11}{3}$

Also, $\alpha\beta = \dfrac{c}{a}$

$= -\dfrac{8}{3}$

Recall that: $(\alpha - \beta)^2 = (\alpha + \beta)^2 - 4\alpha\beta$

$$= (-\frac{11}{3})^2 - 4(-\frac{8}{3})$$
$$= \frac{121}{9} + \frac{32}{3}$$
$$= \frac{121 + 96}{9}$$
$$= \frac{217}{9}$$
$$\therefore (\alpha - \beta)^2 = \frac{217}{9}$$

9. If the roots of the equation $2x^2 - 3x - 1 = 0$ are α and β, find the value of $(\alpha - \beta)^3$

<u>Solution</u>

$2x^2 - 3x - 1 = 0$

From this equation: a = 2, b = –3 and c = –1

But, $\alpha + \beta = -\frac{b}{a}$
$$= -(\frac{-3}{2})$$
$$= \frac{3}{2}$$

Also, $\alpha\beta = \frac{c}{a}$
$$= -\frac{1}{2}$$

Let us first find the value of $\alpha - \beta$ in order to determine $(\alpha - \beta)^3$

Recall that: $\alpha - \beta = \sqrt{(\alpha + \beta)^2 - 4\alpha\beta}$
$$= \sqrt{(\frac{3}{2})^2 - 4(-\frac{1}{2})}$$
$$= \sqrt{\frac{9}{4} + 2}$$
$$= \sqrt{\frac{9 + 8}{4}}$$
$$= \sqrt{\frac{17}{4}}$$
$$\alpha - \beta = 2.0616$$
$$\therefore (\alpha - \beta)^3 = (2.0616)^3$$
$$= 8.76$$

10. If α and β are the roots of the quadratic equation $3x^2 - 5x - 1 = 0$, form an equation whose roots are:

a. $(\alpha^2 + \frac{1}{\beta})$ and $(\beta^2 + \frac{1}{\alpha})$

b. $(\alpha - \frac{1}{\beta})$ and $(\beta - \frac{1}{\alpha})$

Solution

$3x^2 - 5x - 1 = 0$

From this equation: a = 3, b = –5 and c = –1

Hence, $\alpha + \beta = -\frac{b}{a}$

$= -(\frac{-5}{3})$

$= \frac{5}{3}$

Also, $\alpha\beta = \frac{c}{a}$

$= -\frac{1}{3}$

Now, roots of new equation are $(\alpha^2 + \frac{1}{\beta})$ and $(\beta^2 + \frac{1}{\alpha})$

∴ Sum of roots of new equation $= \alpha^2 + \frac{1}{\beta} + \beta^2 + \frac{1}{\alpha}$

$= \alpha^2 + \beta^2 + \frac{1}{\beta} + \frac{1}{\alpha}$ (By rearrangement)

$= (\alpha + \beta)^2 - 2\alpha\beta + \frac{\alpha + \beta}{\alpha\beta}$ (Note that $\alpha^2 + \beta^2 = (\alpha + \beta)^2 - 2\alpha\beta$)

$= (\frac{5}{3})^2 - 2(-\frac{1}{3}) + \frac{\frac{5}{3}}{-\frac{1}{3}}$

$= \frac{25}{9} + \frac{2}{3} - (\frac{5}{3} \times \frac{3}{1})$

$= \frac{25}{9} + \frac{2}{3} - 5$

$= \frac{25 + 6 - 45}{9}$

$= -\frac{14}{9}$

Sum of roots $(\alpha + \beta) = -\frac{14}{9}$

And product of roots of new equation $= (\alpha^2 + \frac{1}{\beta})(\beta^2 + \frac{1}{\alpha})$

$= \alpha^2\beta^2 + \frac{\alpha^2}{\alpha} + \frac{\beta^2}{\beta} + \frac{1}{\alpha\beta}$

$$= (\alpha\beta)^2 + \alpha + \beta + \frac{1}{\alpha\beta}$$

$$= (-\frac{1}{3})^2 + \frac{5}{3} + \frac{1}{-\frac{1}{3}}$$

$$= \frac{1}{9} + \frac{5}{3} - 3$$

$$= \frac{1 + 15 - 27}{9}$$

$$= -\frac{11}{9}$$

Product of roots of new equation $(\alpha\beta) = -\frac{11}{9}$

Recall that a quadratic equation is represented as:

$x^2 - $ (sum of roots)$x + $ (product of roots) $= 0$

Or, $x^2 - (\alpha + \beta)x + (\alpha\beta) = 0$

When the values obtained for $\alpha + \beta$ and $\alpha\beta$ are substituted into the equation above, it gives:

$$x^2 - (-\frac{14}{9})x - \frac{11}{9} = 0$$

When each term is multiplied by 9 in order to clear out the fractions, it gives:

$9x^2 + 14x - 11 = 0$

∴ The equation whose roots are $(\alpha^2 + \frac{1}{\beta})$ and $(\beta^2 + \frac{1}{\alpha})$ is $9x^2 + 14x - 11 = 0$

b. Since roots of new equation are $(\alpha - \frac{1}{\beta})$ and $(\beta - \frac{1}{\alpha})$,

∴ Sum of roots of new equation $= \alpha - \frac{1}{\beta} + \beta - \frac{1}{\alpha}$

$$= \alpha + \beta - (\frac{1}{\beta} + \frac{1}{\alpha}) \quad \text{(By rearrangement)}$$

$$= \alpha + \beta - (\frac{\alpha + \beta}{\alpha\beta})$$

$$= \frac{5}{3} - \frac{\frac{5}{3}}{-\frac{1}{3}}$$

$$= \frac{5}{3} + (\frac{5}{3} \times \frac{3}{1})$$

$$= \frac{5}{3} + 5$$

$$= \frac{5 + 15}{3}$$

Sum of roots $(\alpha + \beta) = \frac{20}{3}$

And product of roots of new equation $= (\alpha - \frac{1}{\beta})(\beta - \frac{1}{\alpha})$

$$= \alpha\beta - \frac{\alpha}{\alpha} - \frac{\beta}{\beta} + \frac{1}{\alpha\beta}$$

$$= \alpha\beta - 1 - 1 + \frac{1}{\alpha\beta}$$

$$= \alpha\beta - 2 + \frac{1}{\alpha\beta}$$

$$= -\frac{1}{3} - 2 + \frac{1}{-\frac{1}{3}}$$

$$= -\frac{1}{3} - 2 - 3$$

$$= -\frac{1}{3} - 5$$

$$= \frac{-1 - 15}{3}$$

$$= -\frac{16}{3}$$

Product of roots of new equation $(\alpha\beta) = -\frac{16}{3}$

Recall that a quadratic equation is represented as:

$x^2 - (\text{sum of roots})x + (\text{product of roots}) = 0$

Or, $x^2 - (\alpha + \beta)x + (\alpha\beta) = 0$

When the values obtained for $\alpha + \beta$ and $\alpha\beta$ are substituted into the equation above, it gives:

$$x^2 - (\frac{20}{3})x - \frac{16}{3} = 0$$

When each term is multiplied by 3 in order to clear out the fractions, it gives:

$3x^2 - 20x - 16 = 0$

∴ The equation whose roots are $(\alpha - \frac{1}{\beta})$ and $(\beta - \frac{1}{\alpha})$ is $3x^2 - 20x - 16 = 0$

Maximum and Minimum Values of a Quadratic Function

The maximum or minimum value of a quadratic function is given by:

$$\frac{4ac - b^2}{4a}$$

The equation of the line of symmetry of a quadratic curve is given by: $x = -\frac{b}{2a}$

This line is called the axis of symmetry of a quadratic curve.

For a quadratic equation to have equal roots, the following condition must be met:

$b^2 = 4ac$ (From $b^2 - 4ac = 0$)

A quadratic equation which has equal roots is also a perfect square. Therefore, if a quadratic equation is a perfect square, then it follows that:

$b^2 = 4ac$

Examples
1. If $(2x + p)(3x - q) = 6x^2 - 7x + 2$, where p and q are constant, find the possible values of p.
Solution
$(2x + p)(3x - q) = 6x^2 - 7x + 2$
Let us expand the left hand side. This gives
$6x^2 - 2qx + 3px - pq = 6x^2 - 7x + 2$
$6x^2 - (2q - 3p)x - pq = 6x^2 - 7x + 2$ [Note: $-2qx + 3px$ has been factorized to give $-(2q - 3p)x$]
Comparing the terms in x on both sides of the equation shows that:
$-(2p - 3p) = -7$ (The terms in x)
∴ $2p - 3p = 7$ (After dividing both sides by -1)
∴ $2p - 3p = 7$ .... Equation (1)
Similarly, comparing the constant terms on both sides of the equation shows that:
   $-pq = 2$
∴ $pq = -2$ (After dividing both sides by -1)
   $pq = -2$ .... Equation(2)
From equation (2):
$q = -\dfrac{2}{p}$
Substitute $-\dfrac{2}{p}$ for q in equation (1)
$2q - 3p = 7$ .... Equation(1)
$2(-\dfrac{2}{p}) - 3p = 7$
$-\dfrac{4}{p} - 3p = 7$
Multiply each term by p in order to clear the fraction.
$p(-\dfrac{4}{p}) - p(3p) = p(7)$
$-4 - 3p^2 = 7p$
$0 = 3p^2 + 7p + 4$
∴ $3p^2 + 7p + 4 = 0$
Solving this equation by factorization method gives:
   $3p^2 + 3p + 4p + 4 = 0$
   $3p(p + 1) + 4(p + 1) = 0$
∴ $(p + 1)(3p + 4) = 0$
Equating each bracket to zero and solving each equation gives $p = -1$ or $p = -\dfrac{4}{3}$
∴ The possible values of p are -1 and $-\dfrac{4}{3}$

2. If $2x^2 - (m - 4)x - 4(m + 2) = 0$ has equal roots, find the possible values of m

Solution

$2x^2 - (m - 4)x - 4(m + 2) = 0$

From the equation, the values of a, b and c are given by $a = 2$, $b = -(m - 4)$, $c = -4(m + 2)$

Or, $a = 2$, $b = -m + 4$, $c = -4m - 8$ (When we expand the brackets above)

For a quadratic equation to have equal roots, the following condition must be met:

$b^2 = 4ac$ (From $b^2 - 4ac = 0$)

Substituting the values above gives:

$b^2 = 4ac$

$(-m + 4)^2 = 4 \times 2 \times (-4m - 8)$ (By substituting the values of a, b and c)

$(4 - m)^2 = 8(-4m - 8)$ (Note that $-m + 4 = 4 - m$)

$(4 - m)(4 - m) = -32m - 64$

$16 - 4m - 4m + m^2 = -32m - 64$

$m^2 - 8m + 32m + 16 + 64 = 0$

$m^2 + 24m + 80 = 0$

∴ $(m + 20)(m + 4) = 0$ (After factorization)

∴ $m = -20$ or $m = -4$

Hence, the possible values of m are -20 and -4.

3. Find the value of the constant k for which the expression $4x^2 + 20x + (12 + k)$ is a perfect square.

Solution

$4x^2 + 20x + (12 + k)$

From this equation, a, b and c are:

$a = 4$, $b = 20$, $c = 12 + k$

For a quadratic equation to be a perfect square, the condition below must be met.

$b^2 = 4ac$

$20^2 = 4 \times 4 \times (12 + k)$

$400 = 16(12 + k)$

$400 = 192 + 16K$

$400 - 192 = 16K$

∴ $k = \dfrac{208}{16}$

$k = 13$

4. Find the maximum value of $5 - 11x - 2x^2$

Solution

Recall that maximum value is given by:

$$\frac{4ac-b^2}{4a}$$

From the equation $5 - 11x - 2x^2$:

$a = -2, b = -11, c = 5$

$\therefore$ Maximum value $= \dfrac{4ac-b^2}{4a}$

$= \dfrac{4(-2 \times 5) - (-11)^2}{4(-2)}$

$= \dfrac{-40 - 121}{-8}$

$= \dfrac{-161}{-8}$

$= \dfrac{161}{8}$

5. Given the function, $y = 3x^2 + 7x - 10$, find:
a. the minimum value of y
b. the value of x for which y is minimum

Solution

a. $y = 3x^2 + 7x - 10$

From the equation:

$a = 3, b = 7, c = -10$

The minimum value of y is given by:

$y = \dfrac{4ac - b^2}{4a}$

$= \dfrac{4(3)(-10) - (7)^2}{4(3)}$

$= \dfrac{-120 - 49}{12}$

$= \dfrac{-169}{12}$

$= -\dfrac{169}{12}$

b. The value of x for which y is minimum is given by:

$x = \dfrac{-b}{2a}$

$= \dfrac{-7}{2(3)}$

$x = \dfrac{-7}{6}$ (Note that this is also the line of symmetry of the quadratic curve)

6. Given the function, $y = 12 - 9x - 2x^2$, find:
a. the maximum value of y
b. the value of x for which y is maximum

Solution

a. $y = 12 - 9x - 2x^2$

From the equation:

$a = -2, b = -9, c = 12$

The maximum value of y is given by:

$$y = \frac{4ac - b^2}{4a}$$

$$= \frac{4(-2)(12) - (-9)^2}{4(-2)}$$

$$= \frac{-96 - 81}{-8}$$

$$= \frac{-177}{-8}$$

$$= \frac{177}{8}$$

b. The value of x for which y is maximum is given by:

$$x = \frac{-b}{2a}$$

$$= \frac{-(-9)}{2(-2)}$$

$$x = \frac{9}{-4} \quad \text{(Note that this is also the line of symmetry of the quadratic curve)}$$

$$x = -\frac{9}{4}$$

7. One root of the quadratic equation $x^2 - (4 + p)x + 12 = 0$ is three times the other. Find:
a. the roots of the equation
b. the possible values of p

Solution

a. $x^2 - (4 + p)x + 12 = 0$

Let one of the roots be α. Therefore, the other root is 3α.

From the equation:

Product of roots = 12   (From $\frac{c}{a}$ in the equation. Note that a = 1)

$\therefore \quad \alpha(3\alpha) = 12$

$3\alpha^2 = 12$

$\alpha^2 = 4$ (After dividing both sides by 3)

∴ α = √4
   α = ±2
∴ α = 2 or –2
When α = 2, the other root is 3α = 3 x 2 = 6
When α = –2, the other root is 3α = 3(–2) = –6
Therefore, the roots are either 2 and 6, or –6 and –2

b. From the equation: $x^2 - (4 + p)x + 12 = 0$:
   a = 1, b = –(4 + p), c = 12
Sum of the roots = –b/a
   α + 3α = – [–(4 + p)]/1   (Note that α and 3α are the two roots)
   4α = 4 + p
   4α – 4 = p
∴ When α = 2, p is given by
   4α – 4 = p
   4(2) – 4 = p
   8 – 4 = p
∴ p = 4
Also, when α = –2, p is given by:
   4α – 4 = p
   4(–2) – 4 = p
   – 8 – 4 = p
   p = –12
∴ The possible values of p are 4 and –12.

Exercise 7

1. If α and β are the roots of the quadratic equation $3x^2 - 8x + 5 = 0$, find:
a. $\alpha + \beta$
b. $\alpha\beta$
c. $\alpha^2 + \beta^2$
d. $\dfrac{\alpha}{\beta} + \dfrac{\beta}{\alpha}$
e. $\dfrac{1}{\alpha} + \dfrac{1}{\beta}$
f. $\dfrac{1}{\alpha^2} + \dfrac{1}{\beta^2}$
g. $\alpha^3 + \beta^3$

2. If the roots of the quadratic equation, $2x^2 + 12x - 11 = 0$, are α and β, find the equation whose

roots are α^2 and β^2

3. If the roots of the quadratic equation $4x^2 + 3x - 10 = 0$, are α and β, find the equation whose roots are $\frac{\alpha}{\beta}$ and $\frac{\beta}{\alpha}$

4. Given that the quadratic equation $x^2 - 15x - 2 = 0$, has roots α and β, find the equation whose roots are $\frac{1}{\alpha}$ and $\frac{1}{\beta}$

5. If α and β are the roots of the quadratic equation $3x^2 - 15x + 7 = 0$, find the quadratic equation whose roots are $\frac{1}{\alpha^2}$ and $\frac{1}{\beta^2}$

6. If α and β are the roots of the quadratic equation $2x^2 + 5x - 6 = 0$, find the quadratic equation whose roots are α^3 and β^3

7. If α and β are the roots of the equation $5x^2 - 8x + 2 = 0$, find the values of:
   a. $\alpha - \beta$
   b. $\alpha^2 - \beta^2$

8. If α and β are the roots of the equation $10x^2 + 15x - 8 = 0$, find the value of $(\alpha - \beta)^2$

9. If the roots of the equation $4x^2 - 2x - 3 = 0$ are α and β, find the value of $(\alpha - \beta)^3$

10. If α and β are the roots of the quadratic equation $x^2 - 2x - 3 = 0$, form an equation whose roots are:
    a. $(\alpha^2 + \frac{1}{\beta})$ and $(\beta^2 + \frac{1}{\alpha})$
    b. $(\alpha - \frac{1}{\beta})$ and $(\beta - \frac{1}{\alpha})$

11. If $(3x + m)(x - n) = 3x^2 - 11x + 6$, where m and n are constants, find the possible values of m and n.

12. If $2x^2 - (p + 3)x - 2(p - 2) = 0$ has equal roots, find the possible values of p.

13. Find the value of the constant k for which the expression $3x^2 + 4x - (8 + k)$ is a perfect square.

14. Find the maximum value of $7 - 14x - 5x^2$

15. Given the function, $y = 2x^2 + 11x - 12$, find:
    a. the minimum value of y
    b. the value of x for which y is minimum

16. Given the function, $y = 18 - 15x - 4x^2$, find:
    a. the maximum value of y
    b. the value of x for which y is maximum

17. One root of the quadratic equation $2x^2 - (2 - k)x + 6 = 0$ is three times the other. Find:
    a. the roots of the equation
    b. the possible values of k

18. If α and β are the roots of the equation $2x^2 - 6x + 3 = 0$, find the values of $\alpha^2 - \beta^2$

19. Find the value of the constant k for which the expression $5x^2 + 10x - (3 + k)$ is a perfect square.

20. One root of the quadratic equation $4x^2 - (9 - k)x + 10 = 0$ is twice the other. Find the roots of the

equation.

CHAPTER 8
FUNCTIONS

A function, f, whose output is y, and input or variable is x, is a rule which describes how a value of x is used to obtain a value of y. A function is usually represented in the form of an equation as follows:
$$y = f(x)$$
The rule of a function must imply that for each input of x, there must be only one output of y. For example, the equation:
$$y = 5x - 1$$
shows that for every value of x, there is only one value of y. Hence this rule is a function.
However, an equation such as:
$$y = \sqrt{x} \quad \text{or} \quad y = x^{\frac{1}{2}}$$
is a rule that can give two possible values of y. If $x = 4$, then $y = 4^{\frac{1}{2}}$. Which gives:
$$y = \sqrt{4}$$
$$y = \pm 2 \text{ which means } +2 \text{ or } -2.$$
This shows that for a positive value of x, there are two possible values of y. Hence, $y = x^{\frac{1}{2}}$ is not a function. A function should give only one value of the output y.

Examples
Determine if each of the following is a function or not.
1. $y = x^{\frac{1}{4}} - 2$
2. $y = 3x^2 - 5\sqrt[3]{x}$
3. $y = \sqrt{20 - x^2}$
4. $y = \dfrac{1}{2x - 1}$

Solutions
1. $y = x^{\frac{1}{4}} - 2$

In order to test if this is a function, we can put a value of x and see if we will get only one value of y. Let us take $x = 16$ and put it in the equation. This gives:
$$y = x^{\frac{1}{4}} - 2$$
$$y = 16^{\frac{1}{4}} - 2$$
$$= \sqrt[4]{16} - 2$$
$$= \pm 2 - 2$$
$$= +2 - 2 \text{ or } -2 - 2$$

Hence, y = 0 or y = −4

Therefore, $y = x^{\frac{1}{4}} - 2$ is not a function since a value of x produces two values of y.

2. $y = 3x^2 - 5\sqrt[3]{x}$

Let us put $x = -8$ into the equation. Note that any convenient value of can be used.

$y = 3x^2 - 5\sqrt[3]{x}$
$ = 3(-8)^2 - 5\sqrt[3]{-8}$
$ = 3(64) - 5(-2)$
$ = 192 + 10$
$ = 202$

Hence, $y = 3x^2 - 5\sqrt[3]{x}$ is a function since only one value of y is produced.

3. $y = \sqrt{20 - x^2}$

Let us put $x = 3$ into the equation. Be careful to put in the square root sign, only values of x that will give a positive value. This is because we cannot evaluate values such as $\sqrt{-10}$, but only positive values such as $\sqrt{10}$

Hence, $y = \sqrt{20 - x^2}$
$ = \sqrt{20 - 3^2}$ (Since $x = 3$ as stated above)
$ = \sqrt{20 - 9}$
$ = \sqrt{11}$
$y = \pm \sqrt{11}$

Therefore, $y = +\sqrt{11}$ or $-\sqrt{11}$

Hence, $y = \sqrt{20 - x^2}$ is not a function since there are two possible values is y.

4. $y = \dfrac{1}{2x - 1}$

Let us put $x = -2$ into the equation as follows:

$y = \dfrac{1}{2x - 1}$
$ = \dfrac{1}{2(-2) - 1}$
$ = \dfrac{1}{-4 - 1}$
$ = \dfrac{1}{-5}$
$y = -0.2$

Hence, $y = \dfrac{1}{2x - 1}$ is a function

Domain and Range of a Function

Domain are all the values of x that a function can take and process.

Range is the set of values y obtained from each domain of a function. Range is also called co-domain.

For example, if $y = \sqrt{4 - x^2}$, where x and y are real numbers, and x is a whole number, then the domain is the set of values –2, –1, 0, 1, 2. These are the values of x that will not give negative values in the square root sign. The range (i.e. the corresponding values of y from the domain values of x) is the set of values 0, √3, 2, √3, 0, or simply 0, √3, 2, when avoiding repetition.

A function can be defined by a given domain of values.

For example if a function is defined as follows:

$y = 2x, \quad -5 < x < 1$

then from the defined values, the domain of the function is:

$-5 < x < 1$

When each of the extreme values (i.e. –5 and 1) is substituted into the function, then the range of the function will be obtained as:

$y = 2(-5) = -10 \quad$ (When $x = -5$)

and $y = 2(1) = 2 \quad$ (When $x = 1$)

This gives the range: $\quad -10 < x < 2$

If a function is given by:

$$y = \frac{1}{(x-2)(x+5)}$$

then the domain can be any value of x except $x = 2$ and $x = -5$. These two values of x will give a denominator of zero which makes the function undefined. Let us see the result of these two values of x. When $x = 2$, then:

$$y = \frac{1}{(x-2)(x+5)}$$
$$= \frac{1}{(2-2)(2+5)}$$
$$= \frac{1}{(0)(7)} = \frac{1}{0} \quad \text{(This is undefined)}$$

When $x = -5$, then:

$$y = \frac{1}{(x-2)(x+5)}$$
$$= \frac{1}{(-5-2)(-5+5)}$$
$$= \frac{1}{(-7)(0)} = \frac{1}{0} \quad \text{(This is undefined)}$$

Hence, $x = 2$ and $x = -5$ are two values that should not be in the domain. Hence, the range of the function should not be calculated when $x = 2$ and $x = -5$.

Arithmetic Operations of Function

Examples

1. If $f(x) = 2x^2 + 1$, find:
   a. $f(-2)$
   b. $f(5)$
   c. $f(0.3)$
   d. $f(x - 1)$
   e. $f(2x + 3)$

Solutions

a. $f(x) = 2x^2 + 1$

In order to obtain f(−2), we simply substitute −2 for x in the given function.

Hence, $f(x) = 2x^2 + 1$

$f(-2) = 2(-2)^2 + 1$

$\qquad = 2(4) + 1 = 8 + 1$

$\qquad = 9$

Therefore, $f(-2) = 9$

b. $f(x) = 2x^2 + 1$

$f(5) = 2(5)^2 + 1$

$\qquad = 2(25) + 1 = 50 + 1$

$\qquad = 51$

Therefore, $f(5) = 51$

c. $f(x) = 2x^2 + 1$

$f(0.3) = 2(0.3)^2 + 1$

$\qquad = 2(0.09) + 1$

$\qquad = 0.18 + 1$

$\qquad = 1.18$

Therefore, $f(0.3) = 1.18$

d. $f(x) = 2x^2 + 1$

$f(x - 1) = 2(x - 1)^2 + 1$

$\qquad = 2(x - 1)(x - 1) + 1$

$\qquad = 2(x^2 - x - x + 1) + 1$

$\qquad = 2(x^2 - 2x + 1) + 1$

$\qquad = 2x^2 - 4x + 2 + 1$

$\qquad = 2x^2 - 4x + 3$

Therefore, $f(x - 1) = 2x^2 - 4x + 3$

e. $f(x) = 2x^2 + 1$
   $f(2x + 3) = 2(2x + 3)^2 + 1$
   $= 2(2x + 3)(2x + 3) + 1$
   $= 2(4x^2 + 6x + 6x + 9) + 1$
   $= 2(4x^2 + 12x + 9) + 1$
   $= 8x^2 + 24x + 18 + 1$
   $= 8x^2 + 24x + 19$

   Therefore, $f(2x + 3) = 8x^2 + 24x + 19$

2. A function f is defined by $f(x) = 5x - 8$
   a. Find $f(\frac{1}{2})$
   b. If $f(3m - 1) = 14$, find the value of m^2
   c. Find $5f(-2)$

Solution

a. $f(x) = 5x - 8$
   $f(\frac{1}{2}) = 5(\frac{1}{2}) - 8$
   $= \frac{5}{2} - 8 = \frac{5 - 16}{2}$
   $= \frac{-11}{2}$

   Therefore, $f(\frac{1}{2}) = -5\frac{1}{2}$

b. $f(x) = 5x - 8$
   $f(3m - 1) = 5(3m - 1) - 8$
   $14 = 15m - 5 - 8$ (Note that $f(3m - 1) = 14$)
   $14 = 15m - 13$
   $14 + 13 = 15m$
   $27 = 15m$
   $m = \frac{27}{15}$
   $m = \frac{9}{5}$

   Therefore, $m^2 = (\frac{9}{5})^2$
   $m^2 = \frac{81}{25}$

c. $f(x) = 5x - 8$
   $f(-2) = 5(-2) - 8$

88

$$= -10 - 8$$
$$f(-2) = -18$$

Therefore, 5f(−2) is obtained by simply multiplying f(−2) by 5.
Therefore, 5f(−2) = 5(−18)
$$= -90$$

3. If $f(x) = 2x + 5$ and $g(x) = x - 2$, find:
a. $f(x) + g(x)$
b. $g(x) - 2f(x)$
c. $f(x) \times g(x)$
d. $f(2) \div g(5)$
e. $h(x) = f(x - 1) - g(-3)$
f. $h(-1)$

Solutions
a. $f(x) + g(x)$
This is the addition of the two functions. It is given as follows:
$$f(x) + g(x) = 2x + 5 + x - 2$$
$$= 3x + 3$$

b. $g(x) - 2f(x)$
In this case, 2f(x) has to be evaluated first. The overall solution is given by:
$$g(x) - 2f(x) = x - 2 - [2(2x + 5)]$$
$$= x - 2 - (4x + 10)$$
$$= x - 2 - 4x - 10$$
$$= -3x - 12$$

c. $f(x) \times g(x) = (2x + 5)(x - 2)$
$$= 2x^2 - 4x + 5x - 10$$
$$= 2x^2 + x - 10$$

d. $f(2) \div g(5)$
Let us determine f(2) and g(5) separately as follows:
$$f(x) = 2x + 5$$
$$f(2) = 2(2) + 5$$
$$= 4 + 5$$
$$= 9$$
$$g(x) = x - 2$$

$$g(5) = 5 - 2$$
$$= 3$$
$$\therefore f(2) \div g(5) = \frac{9}{3}$$
$$= 3$$

e. $h(x) = f(x - 1) - g(-3)$
Let us find $f(x - 1)$ as follows:
$$f(x) = 2x + 5$$
Hence, $f(x - 1) = 2(x - 1) + 5$
$$= 2x - 2 + 5$$
$$f(x - 1) = 2x + 3$$
Also, let us find $g(-3)$ as follows:
$$g(x) = x - 2$$
$$g(-3) = -3 - 2$$
$$g(-3) = -5$$
$\therefore h(x) = f(x - 1) - g(-3)$
$$= 2x + 3 - (-5) \quad \text{(Since } f(x - 1) = 2x + 3 \text{ and } g(-3) = -5\text{)}$$
$$= 2x + 3 + 5$$
$$h(x) = 2x + 8$$

f. $h(x) = 2x + 8$
$$h(-1) = 2(-1) + 8$$
$$= -2 + 8$$
$$h(-1) = 6$$

Composing Functions

Chains of functions can be obtained when two or more functions combine together. For example if $f(x) = 2x$ and $g(x) = x - 1$, then a third function such as $h(x) = g[f(x)]$ can be obtained by the combination of $f(x)$ and $g(x)$. It is written as $h(x) = g \circ f(x)$, and is read as h of x equals g of f of x. h is said to be a function of function. Note that any letter can be used to represent a function.

Examples
1. If $a(x) = 2x - 1$ and $b(x) = -3x$, find
a. $f(x) = a[b(x)]$
b. $g(x) = b[a(x)]$

Solutions
a. $f(x) = a[b(x)]$
This can also be written as: $f(x) = a \circ b(x)$

In order to find f(x), we simply substitute b(x) (i.e. –3x) for x in a(x). This means to put $x = -3x$ in a(x). This gives:

$f(x) = a[b(x)]$
$= a(-3x)$ (Since b(x) = –3x)

But, $a(x) = 2x - 1$

∴ $a(-3x) = 2(-3x) - 1$
$= -6x - 1$

Hence, $f(x) = -6x - 1$ [Since f(x) = a(–3x)]

b. $g(x) = b[a(x)]$

In order to find g(x), we simply substitute a(x) (i.e. 2x – 1) for x in b(x). This means we put $x = 2x - 1$ in b(x). This gives:

$g(x) = b[a(x)]$
$= b(2x - 1)$ (Since a(x) = 2x – 1)

But, $b(x) = -3x$

∴ $b(2x - 1) = -3(2x - 1)$
$= -6x + 3$

Hence, $g(x) = -6x + 3$ [Since g(x) = b(2x – 1)]

2. If $a(x) = x^2$, $b(x) = 5x$ and $c(x) = x - 1$, find:
a. $f(x) = b[a[c(x)]]$
b. $g(x) = c[b[c(x)]]$
c. $h(x) = a[b[b(x)]]$

Solution

a. $f(x) = b[a[c(x)]]$

This can also be written as b o a o c(x).

In order to find f(x), we start from the innermost bracket. Hence, let us first find a[c(x)]. This simply means to substitute c(x) in a(x)

$c(x) = x - 1$ and $a(x) = x^2$.

∴ $a[c(x)] = a(x - 1)$ (Since c(x) = x – 1)

But, $a(x) = x^2$

Hence, $a(x - 1) = (x - 1)^2$
$= (x - 1)(x - 1)$
$= x^2 - x - x + 1$

$a(x - 1) = x^2 - 2x + 1$

∴ $a[c(x)] = x^2 - 2x + 1$ (Since a[c(x)] = a(x – 1))

Let us now determine the final output i.e. b[a[c(x)]]. This means to substitute a[c(x)] in b(x)

$a[c(x)] = x^2 - 2x + 1$ and $b(x) = 5x$
Hence, $b[a[c(x)]] = b(x^2 - 2x + 1)$ (Since $a[c(x)] = x^2 - 2x + 1$)
But, $b(x) = 5x$
Therefore, $b(x^2 - 2x + 1) = 5(x^2 - 2x + 1)$
$= 5x^2 - 10x + 5$
Hence, $b[a[c(x)]] = 5x^2 - 10x + 5$ [Since $b[a[c(x)]] = b(x^2 - 2x + 1)$]
Therefore, $f(x) = 5x^2 - 10x + 5$ (Note that $f(x) = b[a[c(x)]]$)

b. $g(x) = c[b[c(x)]]$
Let us first find $b[c(x)]$. This simply means to substitute $c(x)$ in $b(x)$
$c(x) = x - 1$ and $b(x) = 5x$
∴ $b[c(x)] = b(x - 1)$ (Since $c(x) = x - 1$)
But, $b(x) = 5x$
Hence, $b(x - 1) = 5(x - 1)$
$b(x - 1) = 5x - 5$
∴ $b[c(x)] = 5x - 5$ (Since $b[c(x)] = b(x - 1)$)
Let us now determine the final output i.e. $c[b[c(x)]]$. This means to substitute $b[c(x)]$ in $c(x)$
Hence, $b[c(x)] = 5x - 5$ and $c(x) = x - 1$
Hence, $c[b[c(x)]] = c(5x - 5)$ [Since $b[c(x)] = 5x - 5)$]
But, $c(x) = x - 1$
Therefore, $c(5x - 5) = (5x - 5) - 1$
$c(5x - 5) = 5x - 6$
Hence, $c[b[c(x)]] = 5x - 6$ [Since $c[b[c(x)]] = c(5x - 5)$]
Therefore, $g(x) = 5x - 6$ (Note that $g(x) = c[b[c(x)]]$)

c. $h(x) = a[b[b(x)]]$
Note that $a[b[b(x)]]$ is interpreted as a of b of b of x or a o b o b(x).
Let us first find $b[b(x)]$. This simply means to substitute $b(x)$ in $b(x)$
$b(x) = 5x$
∴ $b[b(x)] = b(5x)$
But, $b(x) = 5x$
Hence, $b(5x) = 5(5x)$
$b(5x) = 25x$
∴ $b[b(x)] = 25x$
Let us now find $a[b[b(x)]]$. This means to substitute $b[b(x)]$ in $a(x)$
Recall that: $b[b(x)] = 25x$ and $a(x) = x^2$
Hence, $a[b[b(x)]] = a(25x)$ [Since $b[b(x)] = 25x$]
But, $a(x) = x^2$

Therefore, $a(25x) = (25x)^2$
$a(25x) = 625x^2$
Hence, $a[b[b(x)]] = 625x^2$ [Since $a[b[b(x)]] = a(25x)$]
Therefore, $h(x) = 625x^2$ (Since $h(x) = a[b[b(x)]]$)

Continuous and Discontinuous Functions

If a graph of a function is drawn without having to take ones hand off the graph paper, then the function is a continuous function. The graph of a continuous function has no sudden jump or break. Sine and cosine functions are continuous functions.

Some graphs of some functions make jumps at a point or some points in the interval. Such functions are called discontinuous functions. $y = \tan x$ is a discontinuous function.

Even Functions

A function $f(x)$ is said to be even if $f(x) = f(-x)$ for all values of x. The graphs of even functions always have their line of symmetry as the y–axis. This means that the vertical line $x = 0$ divide the graph into two equal parts.
$f(x) = x^2$ and $f(x) = \cos x$ are examples of even functions.

Odd Functions

A function $f(x)$ is said to be odd if $f(-x) = -f(x)$ for all values of x. Graphs of odd functions usually pass through the origin as a point of symmetry. This means that the origin, (0, 0) divides the graph into two equal parts. $F(x) = x^3$ and $f(x) = \sin x$ are examples of odd functions.
Note that some functions are neither even nor odd.

Examples
1. Classify the following into even and odd functions:
a. $f(x) = \tan x$
b. $f(x) = x^4$

Solution
a. $f(x) = \tan x$
Let us take a value of x such as 60° and use it to test the function as follows:
$F(60°) = \tan 60$
$F(60°) = 1.732$
Now, find $f(-60°)$ to see if the value obtained will be the same as that of $f(60°)$.
$f(-60°) = \tan -60$
$f(-60°) = -1.732$
Comparing the two results [i.e. $f(60°)$ and $f(-60°)$] shows that $f(60°) = 1.732$, while $f(-60°) = -1.732$. Hence the function is an odd function since:

$F(-x) = -f(x)$, i.e. $f(-60°) = -f(60°)$

Note that once the two values obtained are the same size, but different signs, then the function is an odd function.

b. $f(x) = x^4$

Let us take a value of x such as 2.

Hence, $f(2) = 2^4$

$f(2) = 16$

Now use the negative value of x i.e. –2. This gives:

$f(-2) = -2^4$

$F(-2) = 16$

Therefore, $f(2) = f(-2) = 16$

Since the two values obtained are equal and of the same sign, then the function is an even function.

2. Classify the following into even and odd functions:

a. $3x^5$
b. $\cos^3 x$
c. $4x - 1$

Solutions

a. $3x^5$

Let us use $x = 1$

Hence, $3x^5 = 3(1)^5$

$= 3$

When $x = -1$, we have:

$3x^5 = 3(-1)^5$

$= 3 \times (-1) = -3$ (Note that $(-1)^5 = -1$)

Since the values are the same but have opposite signs, then the function is an odd function.

b. $\cos^3 x$

Let $x = 60°$

$\cos^3 x = (\cos x)^3$

$= (\cos 60)^3$

$= (0.5)^3$

$= 0.125$

When $x = -60°$, we have:

$\cos^3 x = (\cos x)^3$

$= [\cos(-60)]^3$

$= (0.5)^3$

= 0.125

The two values obtained are equal. Therefore, the function is an even function.

c. $4x - 1$

Let $x = 5$

$4x - 1 = 4(5) - 1$

$= 20 - 1$

$= 19$

When $x = -5$, we have

$4x - 1 = 4(-5) - 1$

$= -20 - 1$

$= -21$

The two values obtained are not equal. They are entirely different in values. Hence, the function is neither even nor odd.

Inverse of a Function

The inverse of a function $f(x)$, is another function denoted by $f^{-1}(x)$ which reverses the function $f(x)$. For an inverse $f^{-1}(x)$ that exists, the following is true:

$f[f^{-1}(x)] = f^{-1}[f(x)] = x$

The graph of $f^{-1}(x)$ will be a reflection of $f(x)$ in the line $y = x$.

In order to find the inverse of a function, interchange the positions of x and y, and then make y the subject of the formula.

Examples

1. Find the inverse of the following functions:

a. $f(x) = 5x - 1$
b. $f(x) = 2x^2 + 3$
c. $f(x) = 5x^3$

Solution

a. $f(x) = 5x - 1$

Express the function as $y = f(x)$. This gives:

$y = 5x - 1$

Interchange the positions of x and y. This gives:

$x = 5y - 1$

Now make y the subject of the formula.

$x = 5y - 1$

$x + 1 = 5y$

$y = \dfrac{x + 1}{5}$ (After dividing both sides by 5)

Therefore, the inverse of f(x) is:

$f^{-1}(x) = \dfrac{x+1}{5}$ [Simply by replacing y with $f^{-1}(x)$]

b. $f(x) = 2x^2 + 3$

As solved in (a) above, only three steps are involved. They are:
1. equate the function to y
2. interchange the positions of x and y
3. make y the subject of the formula to obtain the inverse of the function.

Note that step 1 is not needed if the function is already equated to y.

Let us now continue as follows:

$f(x) = 2x^2 + 3$
$y = 2x^2 + 3$
$x = 2y^2 + 3$
$x - 3 = 2y^2$
$y^2 = \dfrac{x-3}{2}$
$y = \sqrt{\dfrac{x-3}{2}}$ (This is the required inverse)

Therefore, the inverse of f(x) is:

$f^{-1}(x) = \sqrt{\dfrac{x-3}{2}}$

c. $f(x) = 5x^3$
$y = 5x^3$
$x = 5y^3$
$\dfrac{x}{5} = y^3$
$y = \sqrt[3]{\dfrac{x}{5}}$ (This is the required inverse)

Therefore, the inverse of f(x) is:

$f^{-1}(x) = \sqrt[3]{\dfrac{x}{5}}$

2. Find the inverse of the following functions:

a. $f(x) = (\dfrac{2x-1}{5})^2$

b. $f(x) = (\dfrac{x+1}{3x-4})^{2/3}$

Solutions

a. $f(x) = (\dfrac{2x-1}{5})^2$

$y = (\dfrac{2x-1}{5})^2$

$x = (\dfrac{2y-1}{5})^2$

Taking the square root of both sides gives:

$\sqrt{x} = \sqrt{(\dfrac{2y-1}{5})^2}$

$\sqrt{x} = \dfrac{2y-1}{5}$

Cross multiply to obtain:

$2y - 1 = 5\sqrt{x}$

$2y = 5\sqrt{x} + 1$

$y = \dfrac{5\sqrt{x}+1}{2}$

Therefore, the inverse of $f(x)$ is:

$f^{-1}(x) = \dfrac{5\sqrt{x}+1}{2}$

b. $f(x) = (\dfrac{x+1}{3x-4})^{2/3}$

$y = (\dfrac{x+1}{3x-4})^{2/3}$

$x = (\dfrac{y+1}{3y-4})^{2/3}$

Raise both sides of the equation to a power of $\dfrac{3}{2}$ i.e. the inverse of $\dfrac{2}{3}$ in order to make the power of $(\dfrac{y+1}{3y-4})^{2/3}$ to become 1. This gives:

$x^{\frac{3}{2}} = ((\dfrac{y+1}{3y-4})^{2/3})^{\frac{3}{2}}$

$x^{\frac{3}{2}} = (\dfrac{y+1}{3y-4})^1$ (Note that the powers are multiplied to give 1, i.e. $\dfrac{3}{2} \times \dfrac{2}{3} = 1$)

$(\sqrt{x})^3 = \dfrac{y+1}{3y-4}$ (Note that from indices, $a^{\frac{3}{2}} = (\sqrt{a})^3 = \sqrt{a^3}$

$\sqrt{x^3} = \dfrac{y+1}{3y-4}$

Cross multiply to obtain:

$\sqrt{x^3}(3y-4) = y + 1$

$3y\sqrt{x^3} - 4\sqrt{x^3} = y + 1$

Collect terms in y to obtain:

$$3y\sqrt{x^3} - y = 1 + 4\sqrt{x^3}$$

Factorizing the left hand side gives:

$$y(3\sqrt{x^3} - 1) = 1 + 4\sqrt{x^3}$$

Divide both sides by $3\sqrt{x^3} - 1$ to obtain y as follows:

$$y = \frac{1 + 4\sqrt{x^3}}{3\sqrt{x^3} - 1} \quad \text{(This is the required inverse)}$$

Therefore, the inverse of $f(x)$ is:

$$f^{-1}(x) = \frac{1 + 4\sqrt{x^3}}{3\sqrt{x^3} - 1}$$

3. If $f(x) = \frac{3x + 1}{x + 5}$, find:

a. $f^{-1}(x)$

b. $f^{-1}(-2)$

Solution

a. $f(x) = \frac{3x + 1}{x + 5}$

$$y = \frac{3x + 1}{x + 5}$$

$$x = \frac{3y + 1}{y + 5}$$

Cross multiply to obtain:

$$3y + 1 = x(y + 5)$$
$$3y + 1 = xy + 5x$$

Collect terms in y on the left hand side of the equation.

$$3y - xy = 5x - 1$$

Factorize the left hand side to obtain:

$$y(3 - x) = 5x - 1$$

Dividing both sides of the equation by $3 - x$ gives y as follows:

$$y = \frac{5x - 1}{3 - x}$$

Or, $y = \frac{-(1 - 5x)}{-(x - 3)}$ (After factorizing by taking -1 as a factor)

$y = \frac{1 - 5x}{x - 3}$ (After the negative signs cancel out)

Hence, $y = \frac{5x - 1}{3 - x}$ or $y = \frac{1 - 5x}{x - 3}$

Therefore, the inverse of $f(x)$ is:

$f^{-1}(x) = \frac{5x - 1}{3 - x}$ or $f^{-1}(x) = \frac{1 - 5x}{x - 3}$ (Note that any two of the inverses is correct)

b. $f^{-1}(x) = \dfrac{5x - 1}{3 - x}$

$f^{-1}(-2) = \dfrac{5(-2) - 1}{3 - (-2)}$

$= \dfrac{-10 - 1}{3 + 2}$

$= \dfrac{-11}{5}$

∴ $f^{-1}(-2) = -2\dfrac{1}{5}$

Further Worked Examples on Functions

1. If $f(x - 1) = x^2 - 4x + 3$, find:
a. $f(x)$
b. $f(3)$

<u>Solution</u>

a. $f(x - 1) = x^2 - 4x + 3$

Let $y = x - 1$

Make x the subject of the formula. This gives:

$x = y + 1$

We now substitute $y + 1$ for x in the function above. This gives:

$F(y) = f(x - 1) = x^2 - 4x + 3$

$= (y + 1)^2 - 4(y + 1) + 3$

$= (y + 1)(y + 1) - 4(y + 1) + 3$

$= y^2 + 2y + 1 - 4y - 4 + 3$

$F(y) = y^2 - 2y$

Now, put $y = x$ into f(y) in order to obtain f(x) as follows:

$F(x) = x^2 - 2x$

b. $F(x) = x^2 - 2x$

$f(3) = (3)^2 - 2(3)$

$= 9 - 6$

$= 3$

2. If $f(2x + 1) = 5x - 3$, find $f(x + 3)$

<u>Solution</u>

$f(2x + 1) = 5x - 3$

Let $y = 2x + 1$

Make x the subject of the formula. This gives:

$$x = \frac{y-1}{2}$$

We now substitute $\frac{y-1}{2}$ for x in the function above. This gives:

$$F(y) = f(2x + 1) = 5x - 3$$
$$= 5\left(\frac{y-1}{2}\right) - 3$$
$$= \frac{5y - 5}{2} - 3$$
$$= \frac{5y - 5 - 6}{2}$$
$$F(y) = \frac{5y - 11}{2}$$

This can now be written as a function of x by substituting x for y

$$F(x) = \frac{5x - 11}{2}$$

Substitute $x + 3$ for x in order to obtain $f(x + 3)$ as follows:

$$F(x) = \frac{5x - 11}{2}$$
$$F(x + 3) = \frac{5(x + 3) - 11}{2}$$
$$= \frac{5x + 15 - 11}{2}$$
$$F(x + 3) = \frac{5x + 4}{2}$$

3. Given that $f(x + 3) = x^2 - 7$, find $f(-1)$

<u>Solution</u>

$$f(x + 3) = x^2 - 7$$

We are going to use a more direct method which is different from that used in examples 1 and 2 above.

$$f(x + 3) = x^2 - 7$$

In order to find $f(-1)$, we simply equate the two terms in bracket and make x the subject of the formula. Hence, we have:

$$f(x + 3) = f(-1)$$
$$(x + 3) = -1$$
$$x = -1 - 3$$
$$x = -4$$

We now substitute -4 for x in the original function in order to obtain $f(-1)$. This gives:

$$F(-1) = f(-4) \text{ in the function of } f(x + 3)$$

Hence, $f(x + 3) = x^2 - 7$

$F(-4) = (-4)^2 - 7$
$ = 16 - 7 = 9$
Therefore, $f(-1) = 9$ [Since $f(-1) = f(-4)$ in the function of $f(x + 3)$]

4. If $f(x + 2) = 2x - 5$, find:
a. $f(x)$
b. $f^{-1}(x)$
c. $f(-3)$
d. $f^{-1}(-3)$
e. $f(x - 3)$

Solution

a. $f(x + 2) = 2x - 5$
In order to find $f(x)$, equate the two terms in each of the brackets
$$f(x + 2) = f(x)$$
Hence, $x + 2 = x$
$ x = x - 2$

Be sure to make the x in the first bracket i.e. from the original function, to be the subject of the formula.

Hence $f(x) = f(x - 2)$ in the function of $f(x + 2)$

Now substitute $x - 2$ for x in the given function.

$F(x + 2) = 2x - 5$
$F(x) = f(x - 2) = 2(x - 2) - 5$ ($x - 2$ has been substituted for x in the function)
$ = 2x - 4 - 5$
$F(x) = 2x - 9$

b. $F(x) = 2x - 9$

Let us find $f^{-1}(x)$ i.e. the inverse of $f(x)$. This is done as follows:

$y = 2x - 9$
$x = 2y - 9$
$x + 9 = 2y$
$y = \dfrac{x + 9}{2}$

Therefore, $f^{-1}(x) = \dfrac{x + 9}{2}$

c. $F(x) = 2x - 9$
$ F(-3) = 2(-3) - 9$
$ = -6 - 9$

$\quad\quad\quad$ F(−3) = −15

d. $f^{-1}(x) = \dfrac{x+9}{2}$

$\quad\quad f^{-1}(-3) = \dfrac{-3+9}{2}$

$\quad\quad\quad\quad\quad = \dfrac{6}{2}$

$\quad\quad f^{-1}(-3) = 3$

e. F(x) = 2x − 9
$\quad\quad$ F(x − 3) = 2(x − 3) − 9
$\quad\quad\quad\quad\quad\quad = 2x − 6 − 9$
$\quad\quad$ F(x − 3) = 2x − 15

5. If f(3x − 1) = 6x + 5, find:
a. f(2x + 5)
b. f(x²)

Solution

a. $\quad$ f(3x − 1) = 6x + 5

$\quad\quad$ f(2x + 5) is obtained as follows:

$\quad\quad$ 3x − 1 = 2x + 5 $\quad\quad$ (By equating only terms in the brackets)

$\quad\quad$ 3x = 2x + 5 + 1

$\quad\quad$ 3x = 2x + 6

$\quad\quad x = \dfrac{2x+6}{3}$

Hence, f(2x + 5) = f($\dfrac{2x+6}{3}$) in the function of f(3x − 1)

Therefore substitute $\dfrac{2x+6}{3}$ for x in the original function above. This gives:

$\quad\quad$ f(3x − 1) = 6x + 5

$\quad\quad$ f(2x + 5) = f($\dfrac{2x+6}{3}$) = 6($\dfrac{2x+6}{3}$) + 5

$\quad\quad\quad\quad\quad\quad = 2(2x + 6) + 5 \quad\quad$ (Note that 6 ÷ 3 = 2)

$\quad\quad\quad\quad\quad\quad = 4x + 12 + 5$

$\quad\quad$ F(2x + 5) = 4x + 17

b. $\quad$ f(3x − 1) = 6x + 5

F(x²) will be obtained from f(3x − 1) as follows

$\quad\quad$ 3x − 1 = x² $\quad\quad$ (By equating only terms in the brackets)

$$3x = x^2 + 1$$
$$x = \frac{x^2 + 1}{3}$$

Hence, $f(x^2) = f(\frac{x^2 + 1}{3})$ in the function of $f(3x - 1)$

Therefore substitute $\frac{x^2 + 1}{3}$ for x in the original function above. This gives:

$$f(3x - 1) = 6x + 5$$
$$f(x^2) = f(\frac{x^2 + 1}{3}) = 6(\frac{x^2 + 1}{3}) + 5$$
$$= 2(x^2 + 1) + 5$$
$$= 2x^2 + 2 + 5$$
$$F(x^2) = 2x^2 + 7$$

Exercise 8

1. Determine if each of the following is a function or not.
  a. $y = x^{\frac{1}{2}} + 5$
  b. $y = 2x^3 - \sqrt[4]{x}$
  c. $y = \sqrt{5 - x^5}$
  d. $y = \frac{3}{5x - 2}$
2. If $f(x) = 3x^3 - 2$, find:
  a. $f(-1)$
  b. $f(2)$
  c. $f(0.5)$
  d. $f(3x - 2)$
  e. $f(x + 3)$
3. A function f is defined by $f(x) = 2x - 3$
  a. Find $f(\frac{1}{5})$
  b. If $f(2p + 3) = 10$, find the value of p
  c. Find $9f(-\frac{1}{5})$
4. If $f(x) = x - 3$ and $g(x) = 2x + 1$, find:
  a. $f(x) + g(x)$
  b. $g(x) - 5f(x)$
  c. $f(x) \times g(x)$
  d. $f(-1) \div g(2)$

e. $h(x) = f(x - 2) - g(5)$
f. $h(2)$
5. If $a(x) = x - 2$ and $b(x) = 2x$, find
   a. $f(x) = b[a(x)]$
   b. $g(x) = a[b(x)]$
6. If $a(x) = 2x^2$, $b(x) = 3x$ and $c(x) = 2x + 1$, find:
   a. $f(x) = a[b[c(x)]]$
   b. $g(x) = b[b[a(x)]]$
   c. $h(x) = c[a[b(x)]]$
7. Classify the following into even and odd functions:
   a. $f(x) = \cos x$
   b. $f(x) = 2x^3$
8. Classify the following into even and odd functions:
   a. $4x^2$
   b. $\cos^2 3x$
   c. $9x - 4$
   d. $\sin^2 x$
9. Find the inverse of the following functions:
   a. $f(x) = 5x - 1$
   b. $f(x) = 9x^3 + 2$
   c. $f(x) = 2x^5$
10. Find the inverse of the following functions:
    a. $f(x) = (\frac{x-1}{2})^3$
    b. $f(x) = (\frac{2x+1}{5x-2})^{1/2}$
11. If $f(x) = \frac{5x-1}{2x+1}$, find:
    a. $f^{-1}(x)$
    b. $f^{-1}(-1)$
12. If $f(x - 2) = 2x^2 - 5x - 7$, find:
    a. $f(x)$
    b. $f(-2)$
13. If $f(5x - 1) = 2x + 7$, find $f(x - 2)$
14. Given that $f(x + 4) = x^2 - 2$, find $f(-3)$
15. If $f(2x + 1) = 4x - 3$, find:
    a. $f(x)$
    b. $f^{-1}(x)$
    c. $f(-2)$

d. $f^{-1}(-1)$
e. $f(2x-1)$
16. If $f(x-3) = 7x + 2$, find:
a. $f(x-5)$
b. $f(2x^2)$

CHAPTER 9
POLYNOMIALS

A polynomial is an expression which is a sum of terms containing a variable or variables whose power starts from one and above.

The highest power of a variable in a polynomial is called the degree of the polynomial.

Addition and Subtraction of Polynomials

When adding or subtracting polynomials, the like terms are added or subtracted as the case may be. Note that like terms are terms whose variables have the same power.

Examples
1. If $A = 2x^3 + 7x^2 - 5$, $B = 5x^3 - 11$, and $C = x^3 + 2x^2 + 5x + 3$, find:
a. $A + B$
b. $C - B$
c. $2B + A$
d. $A - 2C + 3B$
e. $A - B - C$

Solutions

a. $A + B = (2x^3 + 7x^2 - 5) + (5x^3 - 11)$
$= 2x^3 + 5x^3 + 7x^2 - 5 - 11$ (Take note of how like terms are brought together)
$= 7x^3 + 7x^2 - 16$

Take note of the arrangement of terms in ascending order of powers of the variables.

b. $C - B = (x^3 + 2x^2 + 5x + 3) - (5x^3 - 11)$
$= x^3 + 2x^2 + 5x + 3 - 5x^3 + 11$
$= x^3 - 5x^3 + 2x^2 + 5x + 3 + 11$
$= -4x^3 + 2x^2 + 5x + 14$ (Note that x^3 also means $1x^3$)

c. $2B + A = 2(5x^3 - 11) + (2x^3 + 7x^2 - 5)$
$= 10x^3 - 22 + 2x^3 + 7x^2 - 5$
$= 10x^3 + 2x^3 + 7x^2 - 22 - 5$
$= 12x^3 + 7x^2 - 27$

d. $A - 2C + 3B = 2x^3 + 7x^2 - 5 - 2(x^3 + 2x^2 + 5x + 3) + 3(5x^3 - 11)$
$= 2x^3 + 7x^2 - 5 - 2x^3 - 4x^2 - 10x - 6 + 15x^3 - 33$

Take note of how a negative sign outside a bracket changes all the signs in the bracket.

$= 2x^3 - 2x^3 + 15x^3 + 7x^2 - 4x^2 - 10x - 5 - 6 - 33$
$= 15x^3 + 3x^2 - 10x - 44$

e. $A - B - C = (2x^3 + 7x^2 - 5) - (5x^3 - 11) - (x^3 + 2x^2 + 5x + 3)$
$= 2x^3 + 7x^2 - 5 - 5x^3 + 11 - x^3 - 2x^2 - 5x - 3$
$= 2x^3 - 5x^3 - x^3 + 7x^2 - 2x^2 - 5x - 5 + 11 - 3$
$= -4x^3 + 5x^2 - 5x + 3$

2. If $f(x) = 2x^3 - x^2 + 6x - 5$, find:
a. $f(-2)$
b. $f(0)$

Solutions
a. $f(x) = 2x^3 - x^2 + 6x - 5$
In order to find f(−2), simply substitute −2 for x in the given function. This gives:
$F(-2) = 2(-2)^3 - (-2)^2 + 6(-2) - 5$
$F(-2) = 2(-8) - (4) - 12 - 5$
$= -16 - 4 - 12 - 5$
$= -37$

b. $f(x) = 2x^3 - x^2 + 6x - 5$
$f(0) = 2(0)^3 - (0)^2 + 6(0) - 5$
$= 0 - 0 + 0 - 5$
$= -5$

Multiplication of Polynomials

When multiplying polynomials, remember to add the powers of the variable according to the multiplication law of indices.

Examples
1. If $A = 2x^3 - 5x^2 + 3x$ and $B = 3x^3 + x^2 - 7x - 5$, find AB.

Solution
METHOD 1
In order to carry out this multiplication, use each term in A to multiply all the terms in B. The use of bracket in doing this is necessary. Also, remember to carry the sign of each term in A.
∴ $AB = (2x^3 - 5x^2 + 3x)(3x^3 + x^2 - 7x - 5)$
$= 2x^3(3x^3 + x^2 - 7x - 5) - 5x^2(3x^3 + x^2 - 7x - 5) + 3x(3x^3 + x^2 - 7x - 5)$
$= 6x^6 + 2x^5 - 14x^4 - 10x^3 - 15x^5 - 5x^4 + 35x^3 + 25x^2 + 9x^4 + 3x^3 - 21x^2 - 15x$

Take note of the addition of the powers of x. We now bring like terms together, i.e. terms having the same powers of x. After that we add/subtract the like terms. This gives:
$AB = 6x^6 + 2x^5 - 15x^5 - 14x^4 - 5x^4 + 9x^4 - 10x^3 + 35x^3 + 3x^3 + 25x^2 - 21x^2 - 15x$

$$AB = 6x^6 - 13x^5 - 10x^4 + 28x^3 + 4x^2 - 15x$$

METHOD 2

In order to carry out this method, we arrange the terms in columns with the polynomial having the higher number of terms above the one having lower number of terms. B has higher number of terms, so we place B above A. Note that AB = BA. Also, arrange like terms above each other when carrying out the initial arrangement and during the multiplication of terms. Note that we use each term in the lower column to multiply all the terms in the higher column by starting from the right hand side of the arrangement. This means that we work from the right hand side to the left hand side. It is similar to the way we carry out multiplication of numbers.

Let us now multiply A and B as follows:

$$\begin{array}{r} 3x^3 + x^2 - 7x - 5 \\ 2x^3 - 5x^2 + 3x \\ \hline +9x^4 + 3x^3 - 21x^2 - 15x \\ -15x^5 - 5x^4 + 35x^3 + 25x^2 \\ +6x^6 + 2x^5 - 14x^4 - 10x^3 \\ \hline 6x^6 - 13x^5 - 10x^4 + 28x^3 + 4x^2 - 15x \end{array}$$

2. If $M = x^3 - 8x^2 + 2x + 1$ and $N = 5x^3 + 2x^2 - 4x - 7$, find MN.

<u>Solution</u>

METHOD 1

$MN = (x^3 - 8x^2 + 2x + 1)(5x^3 + 2x^2 - 4x - 7)$
$= x^3(5x^3 + 2x^2 - 4x - 7) - 8x^2(5x^3 + 2x^2 - 4x - 7) + 2x(5x^3 + 2x^2 - 4x - 7) + 1(5x^3 + 2x^2 - 4x - 7)$
$= 5x^6 + 2x^5 - 4x^4 - 7x^3 - 40x^5 - 16x^4 + 32x^3 + 56x^2 + 10x^4 + 4x^3 - 8x^2 - 14x + 5x^3 + 2x^2 - 4x - 7$
$= 5x^6 + 2x^5 - 40x^5 - 4x^4 - 16x^4 + 10x^4 - 7x^3 + 32x^3 + 4x^3 + 5x^3 + 56x^2 - 8x^2 + 2x^2 - 14x - 4x - 7$
$MN = 5x^6 - 38x^5 - 10x^4 + 34x^3 + 50x^2 - 18x - 7$

METHOD 2

Ensure you multiply the signs of any two terms multiplied together. The working is as shown below.

$$\begin{array}{r} x^3 - 8x^2 + 2x + 1 \\ 5x^3 + 2x^2 - 4x - 7 \\ \hline -7x^3 + 56x^2 - 14x - 7 \\ -4x^4 + 32x^3 - 8x^2 - 4x \\ 2x^5 - 16x^4 + 4x^3 + 2x^2 \\ +5x^6 - 40x^5 + 10x^4 + 5x^3 \\ \hline 5x^6 - 38x^5 - 10x^4 + 34x^3 + 50x^2 - 18x - 7 \end{array}$$

3. Given that $f(x) = 2x^3 - 3$ and $g(x) = 3x^3 - 2x^2 + x - 5$, find $f(x) \cdot g(x)$

Solution

METHOD 1

$f(x) \times g(x) = (2x^3 - 3)(3x^3 - 2x^2 + x - 5)$
$= 2x^3(3x^3 - 2x^2 + x - 5) - 3(3x^3 - 2x^2 + x - 5)$
$= 6x^6 - 4x^5 + 2x^4 - 10x^3 - 9x^3 + 6x^2 - 3x + 15$
$= 6x^6 - 4x^5 + 2x^4 - 19x^3 + 6x^2 - 3x + 15$

METHOD 2

Since $g(x)$ has more terms, we will place $g(x)$ above $f(x)$. Also arrange like terms above each other as shown below

$$\begin{array}{r} 3x^3 - 2x^2 + x - 5 \\ 2x^3 - 3 \\ \hline -9x^3 + 6x^2 - 3x + 15 \\ +6x^6 - 4x^5 + 2x^4 - 10x^3 \\ \hline 6x^6 - 4x^5 + 2x^4 - 19x^3 + 6x^2 - 3x + 15 \end{array}$$

Division of Polynomials

If we divide $2x^3 - 2x^2 + x - 5$ by $x - 5$, then $2x^3 - 2x^2 + x - 5$ is called dividend, while $x - 5$ is called the divisor. The result obtained after the division is called the quotient, while what is left at the end of the division is called the remainder.

Examples

1. Divide $2x^2 + 3x - 5$ by $x - 1$

Solution

The layout is as shown below. Note that 'like terms' are arranged in columns above each other when carrying out the division.

STEP 1: Divide the first term of the dividend by the first term of the divisor. This gives: $\dfrac{2x^2}{x} = 2x$. Then write the $2x$ at the top of the division sign as shown below.

$$\begin{array}{r} 2x \\ x-1{\overline{\smash{\big)}\,2x^2 + 3x - 5}} \end{array}$$

STEP 2: Use the $2x$ obtained above to multiply $x - 1$ and write your answer under the dividend. Note that $2x(x - 1)$ will give $2x^2 - 2x$. This is now written below the corresponding like terms of the dividend as shown below.

$$\begin{array}{r} 2x \\ x-1{\overline{\smash{\big)}\,2x^2 + 3x - 5}} \\ 2x^2 - 2x \end{array}$$

STEP 3: Subtract the like terms as arranged above. This means: $2x^2 - 2x^2 = 0x^2$, while $3x - (-2x) = 3x + 2x = 5x$. We now write $5x$ under its corresponding column and ignore the zero under $2x^2$ since there is no need of writing $0x^2$. This is as shown below.

$$\begin{array}{r} 2x \\ x-1\overline{)2x^2 + 3x - 5} \\ \underline{2x^2 - 2x} \\ 5x \end{array}$$

STEP 4: Bring down the next term of the dividend which is –5. This is as shown below.

$$\begin{array}{r} 2x \\ x-1\overline{)2x^2 + 3x - 5} \\ \underline{2x^2 - 2x} \\ 5x - 5 \end{array}$$

STEP 5: After bringing down –5, we now have a new dividend of $5x - 5$. Use this new dividend to repeat step 1 to step 4 above. This means:

Step 1: $\dfrac{5x}{x} = 5$. Write this as +5 at the top of the division sign.

Step 2: Use the 5 above to multiply $x - 1$. This gives $5x - 5$. Write this below the corresponding like terms of our new dividend which is also $5x - 5$.

Step 3: Subtract the like terms arranged above. This will give zero since $5x - 5$ subtracted from $5x - 5$ gives zero.

Step 4: When your subtraction in step 3 gives zero, and there is no more term to bring down from the original dividend, then you have arrived at your answer. These steps give the final solution as shown below:

$$\begin{array}{r} 2x + 5 \\ x-1\overline{)2x^2 + 3x - 5} \\ \underline{2x^2 - 2x} \\ 5x - 5 \\ \underline{5x - 5} \\ - - - \end{array}$$

$\therefore$ $(2x^2 + 3x - 5) \div (x - 1) = 2x + 5$

Check your answer by multiplying $(x - 1)$ by $(2x + 5)$. It will give $2x^2 + 3x - 5$, which is the dividend. Notice that the steps from 1 to 4 above can be formulated into an acronym written as: DMSBd. D means divide, M means multiply, S means subtract, while Bd means bring down. This acronym can always remind you of the next step to take.

2. Divide $5x^3 + x^2 - 8x - 4$ by $x + 1$

Solution

The question can also be written as:
$$\frac{5x^3 + x^2 - 8x - 4}{x + 1}$$
The workings are explained below.

STEP 1: Divide the first term of the dividend by the first term of the divisor. This gives: $\frac{5x^3}{x} = 5x^2$. Then write the $5x^2$ at the top of the division sign as shown below.

$$\begin{array}{r} 5x^2 \\ x+1 \overline{\smash{)}5x^3 + x^2 - 8x - 4} \end{array}$$

STEP 2: Use the $5x^2$ obtained above to multiply $x + 1$ and write your answer under the dividend. Note that $5x^2(x + 1)$ will give $5x^3 + 5x^2$. This is now written below the corresponding like terms of the dividend as shown below.

$$\begin{array}{r} 5x^2 \\ x+1 \overline{\smash{)}5x^3 + x^2 - 8x - 4} \\ 5x^3 + 5x^2 \end{array}$$

STEP 3: Subtract the like terms as arranged above. This means: $5x^3 - 5x^3 = 0x^3$, while $x^2 - (+5x^2) = x^2 - 5x^2 = -4x^2$. We now write $-4x^2$ under its corresponding column and ignore the zero under $5x^3$ since there is no need of writing $0x^3$. This is as shown below.

$$\begin{array}{r} 5x^2 \\ x+1 \overline{\smash{)}5x^3 + x^2 - 8x - 4} \\ \underline{5x^3 + 5x^2} \\ -4x^2 \end{array}$$

STEP 4: Bring down the next term of the dividend which is $-8x$. This is as shown below.

$$\begin{array}{r} 5x^2 \\ x+1 \overline{\smash{)}5x^3 + x^2 - 8x - 4} \\ \underline{5x^3 + 5x^2} \\ -4x^2 - 8x \end{array}$$

STEP 5: After bringing down $-8x$, we now have a new dividend of $-4x^2 - 8x$ as shown above. Use this new dividend to repeat step 1 to step 4 above. This means:

Step 1: $\frac{-4x^2}{x} = -4x$. Write this at the top of the division sign as shown below.

Step 2: Use the $-4x$ above to multiply $x + 1$. This gives $-4x^2 - 4x$. Write this below the corresponding like terms of our new dividend.

Step 3: Subtract the like terms arranged from step 2 above. This will give $-4x$.

Step 4: Bring down the next term of the original dividend which is -4.

Step 5: After bringing down -4, we now have a new dividend of $-4x - 4$. Use this new dividend to repeat step 1 to step 4 above.

Now let us use the acronym DMSBd to complete the remaining part of the division as follows.

D: $-4x \div x = -4$ (This is written on the division sign)
M: $-4(x + 1) = -4x - 4$ (Write this under $-4x - 4$ which is our present dividend)
S: $-4x - 4 - (-4x - 4)$ will give zero.
Bd: There is nothing more to bring down. Hence our division is complete and we now have our final answer.

All the division processes explained above are as shown below.

$$\begin{array}{r}
5x^2 - 4x - 4 \\
x + 1 \overline{\smash{)}5x^3 + x^2 - 8x - 4} \\
\underline{5x^3 + 5x^2} \\
-4x^2 - 8x \\
\underline{-4x^2 - 4x} \\
-4x - 4 \\
\underline{-4x - 4} \\
- - -
\end{array}$$

$\therefore$ $(5x^3 + x^2 - 8x - 4) \div (x + 1) = 5x^2 - 4x - 4$

Check your answer by multiplying $(x + 1)$ by $(5x^2 - 4x - 4)$. It will give $5x^3 + x^2 - 8x - 4$, which is the dividend.

Points to note as a reminder:

D: During division, only the first term of the dividend is used to divide the first term of the divisor.
M: During multiplication, the answer obtained during division is used to multiply all the terms in the divisor.
S: During subtraction, only the like terms are subtracted. The sign of each term must be carried along with it. Be careful of the negative sign of terms, and the subtraction sign used to carry out the operation.
Bd: Bring down the next term along with its sign. When there is nothing else to bring down, then the division process has ended and the terms on the division sign become the answer.

3. Evaluate $\dfrac{x^3 - 7x - 6}{x - 3}$

Solution
A close look at the dividend shows that there is no term in x^2. Therefore, in order to avoid mistake in our division, it is advisable to include $0x^2$ at the right position in the dividend. Hence the question can be re–written as:

$$\dfrac{x^3 - 0x^2 - 7x - 6}{x - 3}$$

We now set out our division as shown below.

$$\begin{array}{r}x^2+3x+2\\x-3\overline{\smash{)}x^3+0x^2-7x-6}\\\underline{x^3-3x^2}\\3x^2-7x\\\underline{3x^2-9x}\\2x-6\\\underline{2x-6}\\---\end{array}$$

WORKING

D: $\dfrac{x^3}{x} = x^2$ (This is written on the division sign)

M: $x^2(x-3) = x^3 - 3x^2$ (Write this under $x^3 + 0x^2$)

S: $0x^2 - (-3x^2) = 0x^2 + 3x^2 = 3x^2$.

Bd: Bring down $-7x$ to obtain $3x^2 - 7x$ as the new dividend.

We now repeat the process using $3x^2 - 7x$

D: $\dfrac{3x^2}{x} = 3x$ (This is written as $+3x$ on the division sign)

M: $3x(x-3) = 3x^2 - 9x$ (Write this under $3x^2 - 7x$)

S: $3x^2 - 7x - (3x^2 - 9x) = 2x$.

Bd: Bring down -6 to meet $2x$. This gives $2x - 6$ as the new dividend.

Finally, we repeat the process by using $2x - 6$ as dividend.

D: $\dfrac{2x}{x} = 2$ (This is written as $+2$ on the division sign)

M: $2(x-3) = 2x - 6$ (Write this under $2x - 6$)

S: Their subtraction gives zero

Bd: Nothing more to bring down. Hence we have our answer as $x^2 + 3x + 2$.

Therefore, $\dfrac{x^3 - 7x - 6}{x - 3} = x^2 + 3x + 2$

4. Divide $2x^3 - 11x^2y + 3xy^2 + y^3$ by $2x - y$

Solution

$$\begin{array}{r}x^2-5xy-y^2\\2x-y\overline{\smash{)}2x^3-11x^2y+3xy^2+y^3}\\\underline{2x^3-x^2y}\\-10x^2y+3xy^2\\\underline{-10x^2y+5xy^2}\\-2xy^2+y^3\\\underline{-2xy^2+y^3}\\---\end{array}$$

5. Simplify: $\dfrac{x^3 - y^3}{x - y}$

Solution

A careful look at the dividend shows that some terms are not present. That is, they are zero. The missing terms are x^2y and xy^2. One method of knowing the missing term is to raise the divisor to the highest power (degree) of the dividend. This means that if $(x - y)^3$ is evaluated, you will see the missing terms.

Hence we represent the missing terms by $0x^2y$ and $0xy^2$. We now carry out the division as shown below.

$$
\begin{array}{r}
x^2 + xy + y^2 \\
x - y \overline{\smash{)}\, x^3 + 0x^2y + 0xy^2 - y^3} \\
\underline{x^3 - x^2y} \\
x^2y + 0xy^2 \\
\underline{x^2y - xy^2} \\
xy^2 - y^3 \\
\underline{xy^2 - y^3} \\
- \ - \ -
\end{array}
$$

Hence, $\dfrac{x^3 - y^3}{x - y} = x^2 + xy + y^2$

6. Find the quotient and the remainder when $4x^3 - 6x^2 + 8x - 5$ is divided by $2x + 1$

Solution

This is a case where we will have a remainder. In the course of our division, whenever we carry out a subtraction, and there is nothing else to bring down, whatever is left becomes the remainder. Let us now carry out the division as follows.

$$
\begin{array}{r}
2x^2 - 4x + 6 \\
2x + 1 \overline{\smash{)}\, 4x^3 - 6x^2 + 8x - 5} \\
\underline{4x^3 + 2x^2} \\
-8x^2 + 8x \\
\underline{-8x^2 - 4x} \\
12x - 5 \\
\underline{12x + 6} \\
-11
\end{array}
$$

There is nothing else left to bring down. This ends the division.

∴ $(4x^3 - 6x^2 + 8x - 5) \div (2x + 1) = 2x^2 - 4x + 6$, remainder -11.
Hence, the quotient is $2x^2 - 4x + 6$, while the remainder is -11.
This division can be written as: $\dfrac{4x^3 - 6x^2 + 8x - 5}{2x + 1} = 2x^2 - 4x + 6 - \dfrac{11}{2x + 1}$

7. Find the quotient and the remainder when $3x^3 - 7x^2 - x + 9$ is divided by $x - 5$
Solution
Let us carry out the division as follows.

$$\begin{array}{r}
3x^2 + 8x + 39 \\
x - 5 \overline{) 3x^3 - 7x^2 - x + 9} \\
\underline{3x^3 - 15x^2} \\
8x^2 - x \\
\underline{8x^2 - 40x} \\
39x + 9 \\
\underline{39x - 195} \\
204
\end{array}$$

There is nothing else left to bring down. This ends the division.
∴ The quotient is $3x^2 + 8x + 39$, while the remainder is 204.
This division can be written as: $\dfrac{3x^3 - 7x^2 - x + 9}{x - 5} = 3x^2 + 8x + 39 - \dfrac{204}{x - 5}$

Zeros of Polynomials
A zero of a function $f(x)$ is the root of the equation $f(x) = 0$.

Examples
Find the zeros of the following polynomial functions:
1. $f(x) = x^2 - 5x + 6$
2. $f(x) = 2x^2 + 5x - 3$

Solution
1. In order to find the zeros of the function, we equate the function to zero and solve for x as follows:
   $x^2 - 5x + 6 = 0$
   $(x - 2)(x - 3) = 0$
   ∴ $x = 2$ or $x = 3$
Hence, the zeros of the function are 2 and 3.

2. $2x^2 + 5x - 3 = 0$

Solving this equation by factorization gives:
$$2x^2 - x + 6x - 3 = 0$$
$$x(2x - 1) + 3(2x - 1) = 0$$
$$(2x - 1)(x + 3)$$
$$\therefore x = \frac{1}{2} \text{ or } x = -3$$

Hence, the zeros of the function are $\frac{1}{2}$ and -3.

The Factor Theorem

Let us solve the quadratic equation below by factorization.
$$x^2 - 8x - 20 = 0$$
$$(x - 10)(x + 2) = 0$$
$$\therefore x = 10 \text{ or } x = -2$$

This shows that the factor $(x - 10)$ gives a root of 10, while the factor $(x + 2)$ gives a root of -2. This is what the factor theorem means.

Therefore the factor theorem state that:

If $x = a$ is a root of the equation $f(x) = 0$, then $(x - a)$ is a factor of $f(x)$. This also means that $f(a) = 0$ (i.e. substituting 'a' for x in the function gives zero).

The Remainder Theorem

The remainder theorem states that if $f(x)$ is divided by $(x - a)$, the remainder is equal to $f(a)$. Generally, we can say that, if $f(x)$ is divided by $ax - b$, the remainder is equal to $f(\frac{b}{a})$.

Example

1. Factorize $x^3 - 2x^2 - x + 2$ and use it to solve the cubic equation $x^3 - 2x^2 - x + 2 = 0$

Solution

Let us represent the expression as a function of x.
$$F(x) = x^3 - 2x^2 - x + 2$$

We have to employ the method of trial and error to obtain one factor of $f(x)$. Try numbers such as 1, -1, 2, -2, 3 etc.
$$F(x) = x^3 - 2x^2 - x + 2$$
If $x = 1$, then $f(1) = (1)^3 - 2(1)^2 - (1) + 2$
$$= 1 - 2 - 1 + 2$$
$$= 0$$

Since $f(1) = 0$, then $(x - 1)$ is a factor of $f(x)$ (factor theorem). Note that $f(1) = 0$ means that $x = 1$, which can be rearranged to give $x - 1 = 0$ (by taking 1 to the left hand side of the equation). Hence we can see that $x - 1$ is a factor of $f(x)$.

Let us now use the polynomial to divide $(x - 1)$ in order to obtain the other factors. This is shown below.

$$\begin{array}{r} x^2 - x - 2 \\ x - 1 \overline{\smash{)}x^3 - 2x^2 - x + 2} \\ \underline{x^3 - x^2} \\ -x^2 - x \\ \underline{-x^2 + x} \\ -2x + 2 \\ \underline{-2x + 2} \\ - - - \end{array}$$

Hence, $\dfrac{x^3 - 2x^2 - x + 2}{x - 1} = x^2 - x - 2$

Or, $x^3 - 2x^2 - x + 2 = (x - 1)(x^2 - x - 2)$

The quadratic part $x^2 - x - 2$ can be factorized to give:

$x^2 - x - 2 = (x - 2)(x + 1)$

$\therefore \; x^3 - 2x^2 - x + 2 = (x - 1)(x - 2)(x + 1)$

Let us now use the factorized expression above to solve the given equation as follows:

$x^3 - 2x^2 - x + 2 = 0$

$(x - 1)(x - 2)(x + 1) = 0$

Hence, $x = 1$ or 2 or -1 (When each bracket is equated to zero and solved)

2. Solve the equation $2x^3 + 13x^2 + 13x - 10 = 0$

Solution

Let $f(x) = 2x^3 + 13x^2 + 13x - 10$

Let us put numbers such as $-1, 1, -2, 2, -3$ etc, into the function in order to obtain one of the factors.

Hence, if $x = 1$, $f(1) = 2(1)^3 + 13(1)^2 + 13(1) - 10$

$ = 2 + 13 + 13 - 10$

$ = 18$

Therefore, $(x - 1)$ is not a factor

If $x = -1$, $f(-1) = 2(-1)^3 + 13(-1)^2 + 13(-1) - 10$

$ = -2 + 13 - 13 - 10$

$ = -12$

Therefore, $(x + 1)$ is not a factor. Note that from $x = -1$, we have $x + 1$ by taking -1 to the left hand side.

If $x = 2$, $f(2) = 2(2)^3 + 13(2)^2 + 13(2) - 10$

$ = 16 + 52 + 26 - 10$

$ = 84$

Therefore, $(x - 2)$ is not a factor.

If $x = -2$, $f(-2) = 2(-2)^3 + 13(-2)^2 + 13(-2) - 10$
$= -16 + 52 - 26 - 10$
$= 0$

Therefore, $(x + 2)$ is a factor of $f(x)$

We now divide $f(x)$ by $(x + 2)$ in order to get the other factors. This gives:

$$\begin{array}{r}
2x^2 + 9x - 5 \\
x+2\overline{)2x^3 + 13x^2 + 13x - 10} \\
\underline{2x^3 + 4x^2} \\
9x^2 + 13x \\
\underline{9x^2 + 18x} \\
-5x - 10 \\
\underline{-5x - 10} \\
-\ -\ -
\end{array}$$

Hence, $\dfrac{2x^3 + 13x^2 + 13x - 10}{x + 2} = 2x^2 + 9x - 5$

Or, $2x^3 + 13x^2 + 13x - 10 = (x + 2)(2x^2 + 9x - 5)$

The quadratic part $2x^2 + 9x - 5$ can be factorized to give:
$2x^2 + 9x - 5 = 2x^2 + 10x - x - 5$
$= 2x(x + 5) - 1(x + 5)$
$= (x + 5)(2x - 1)$

∴ $2x^3 + 13x^2 + 13x - 10 = (x + 2)(x + 5)(2x - 1)$

Let us now use the factorized expression above to solve the given equation as follows:
$2x^3 + 13x^2 + 13x - 10 = 0$
$(x + 2)(x + 5)(2x - 1) = 0$

Hence, $x = -2$ or -5 or $\dfrac{1}{2}$ (When each bracket is equated to zero and solved)

3. Find the remainder when $x^3 - 2x^2 + 5x + 9$ is divided by $x + 2$.

<u>Solution</u>

Let $f(x) = x^3 - 2x^2 + 5x + 9$

According to the remainder theorem, when this polynomial is divided by $x + 2$, the remainder is obtained from $f(-2)$. Note that -2 is obtained when we set $x + 2 = 0$ and solve it to get $x = -2$.

∴ $f(x) = x^3 - 2x^2 + 5x + 9$
$f(-2) = (-2)^3 - 2(-2)^2 + 5(-2) + 9$
$= -8 - 8 - 10 + 9$
$= -17$

Hence the remainder is -17

4. Find the remainder when $2x^2 + 8x - 3$ is divided by $3x - 1$

Solution

Let $f(x) = 2x^2 + 8x - 3$

According to the remainder theorem, when this polynomial is divided by $3x - 1$, the remainder is obtained from $f(\frac{1}{3})$. Note that $\frac{1}{3}$ is obtained when we set $3x - 1 = 0$ and solve it to get $x = \frac{1}{3}$.

$\therefore \quad f(x) = 2x^2 + 8x - 3$

$f(\frac{1}{3}) = 2(\frac{1}{3})^2 + 8(\frac{1}{3}) - 3$

$= 2(\frac{1}{9}) + \frac{8}{3} - 3$

$= \frac{2}{9} + \frac{8}{3} - 3$

$= \frac{2 + 24 - 27}{9}$

$= -\frac{1}{9}$

Hence the remainder is $-\frac{1}{9}$

5. If $(x + 1)$ and $(3x + 2)$ are factors of $3x^3 + 2x^2 - 3x - 2$, find the third factor.

Solution

Since $(x + 1)$ and $(3x + 2)$ are factors of $3x^3 + 2x^2 - 3x - 2$, then the third factor can be obtained by dividing $3x^3 + 2x^2 - 3x - 2$ by $(x + 1)(3x + 2)$. This is the simple logic:

$(x + 1)(3x + 2)(\quad) = 3x^3 + 2x^2 - 3x - 2$ (Where ( ) is the third factor)

Hence, $(\quad) = \dfrac{3x^3 + 2x^2 - 3x - 2}{(x + 1)(3x + 2)}$ (When both sides of the equation are divide by $(x + 1)(3x + 2)$

$(\quad) = \dfrac{3x^3 + 2x^2 - 3x - 2}{3x^2 + 5x + 2}$ (After expanding the denominator)

We now carry out the division as shown below

$$\begin{array}{r} x - 1 \\ 3x^2 + 5x + 2 \overline{) 3x^3 + 2x^2 - 3x - 2} \\ \underline{3x^3 + 5x^2 + 2x} \\ -3x^2 - 5x - 2 \\ \underline{-3x^2 - 5x - 2} \\ - \quad - \quad - \quad - \end{array}$$

Hence the third factor is $x - 1$

6. If $(3x - 1)$ is a factor of the polynomial $f(x) = 6x^2 + kx - 1$, where k is a constant, find the zeros of $f(x)$.

Solution

Since $3x - 1$ is a factor, then we obtain x as follows:

$3x - 1 = 0$

$3x = 1$

$x = \dfrac{1}{3}$

Hence, $f(\dfrac{1}{3}) = 0$

$f(x) = 6x^2 + kx - 1$

$f(\dfrac{1}{3}) = 6(\dfrac{1}{3})^2 + k(\dfrac{1}{3}) - 1$

$= 6(\dfrac{1}{9}) + \dfrac{k}{3} - 1$

$= \dfrac{6}{9} + \dfrac{k}{3} - 1$

$= \dfrac{2}{3} + \dfrac{k}{3} - 1$

$= \dfrac{2 + k - 3}{3}$

$= \dfrac{k - 1}{3}$

But $f(\dfrac{1}{3}) = 0$

Therefore, $\dfrac{k-1}{3} = 0$

$k - 1 = 3(0)$

$k - 1 = 0$

$k = 1$

Hence the polynomial is $f(x) = 6x^2 + x - 1$ (Since k = 1)

In order to obtain the zeros of the polynomial, we solve the polynomial equation by factorization as follows:

$6x^2 + x - 1 = 0$

$6x^2 + 3x - 2x - 1 = 0$

$3x(2x + 1) - 1(2x + 1) = 0$

$(2x + 1)(3x - 1) = 0$

Therefore, $x = -\dfrac{1}{2}$ or $x = \dfrac{1}{3}$ (i.e. the roots of the equation)

Hence the zeros of f(x) are $-\dfrac{1}{2}$ and $x = \dfrac{1}{3}$

7. Given that $x + 2$ is a factor of the polynomial $2x^3 - x^2 - 7x + 6$, find the other two factors.

<u>Solution</u>

Let us first divide $2x^2 - x^2 - 7x + 6$ by $x + 2$. This is as shown below.

$$\begin{array}{r}
2x^2 - 5x + 3\\
x+2{\overline{\smash{\big)}\,2x^3 - x^2 - 7x + 6}}\\
\underline{2x^3 + 4x^2}\\
-5x^2 - 7x\\
\underline{-5x^2 - 10x}\\
3x + 6\\
\underline{3x + 6}\\
-
\end{array}$$

Hence the quadratic factor of the polynomial is $2x^2 - 5x + 3$. Let us factorize this quadratic expression in order to find the other two linear factors as follows:

$$2x^2 - 5x + 3 = 2x^2 - 3x - 2x + 3$$
$$= x(2x - 3) - 1(2x - 3)$$
$$= (2x - 3)(x - 1)$$

Therefore the other two factors are $(2x - 3)$ and $(x - 1)$

8. The remainder when the polynomial $f(x) = ax^3 + bx^2 + x - 5$ is divided by $x + 2$ is -39, and when it is divided by $x - 1$ the remainder is -3. Determine the values of a and b.

Solution

$f(x) = ax^3 + bx^2 + x - 5$

When $f(x)$ is divided by $x + 2$, then $x = -2$ (When we set $x + 2 = 0$, then $x = -2$)

Hence, $f(-2) = a(-2)^3 + b(-2)^2 + (-2) - 5$

$-39 = a(-8) + b(4) - 2 - 5$ (Note that -39 is the remainder)

$-39 = -8a + 4b - 7$

$8a - 4b = 39 - 7$

$8a - 4b = 32$

$2a - b = 8$ (After dividing each term by 8)

$2a - b = 8$ ....Equation (1)

Similarly, when $f(x)$ is divided by $x - 1$, then $x = 1$ (When we set $x - 1 = 0$, then $x = 1$)

Hence, $f(1) = a(1)^3 + b(1)^2 + (1) - 5$

$-3 = a(1) + b(1) + 1 - 5$ (Note that -3 is the remainder)

$-3 = a + b - 4$

$4 - 3 = a + b$

$a + b = 1$ ....Equation (2)

From equation (2), $a = 1 - b$ ....Equation (3)

Substitute $1 - b$ for a in equation (1) as follows.

$2a - b = 8$ ....Equation (1)

$2(1 - b) - b = 8$

$2 - 2b - b = 8$

$2 - 3b = 8$

$-3b = 8 - 2$

$-3b = 6$

$b = \dfrac{6}{-3}$

$b = -2$

Substitute −2 for b in equation (3) in order to find a.

$a = 1 - b$ ....Equation (3)

$= 1 - (-2)$

$= 1 + 2$

$a = 3$

Therefore, a = 3 and b = −2

9. The polynomial $x^3 + 2x^2 + mx + n$ is divisible by $x + 1$. It leaves a remainder of 12 when it is divided by $x - 2$.

a. Find m and n
b. Factorize the polynomial completely
c. Find the zeros of the polynomial

Solution

a. $x^3 + 2x^2 + mx + n$

Since the polynomial is divisible by $x + 1$, it means that $x + 1$ is a factor of the polynomial. Hence, $x = -1$ is a root of the polynomial equation. Hence substituting $x = -1$, gives zero as follows.

$x^3 + 2x^2 + mx + n$

$(-1)^3 + 2(-1)^2 + m(-1) + n = 0$

$-1 + 2 - m + n = 0$

$-m + n = -1$ ....Equation (1)

Also, since the polynomial leaves a remainder of 12 when divided by $(x - 2)$, it means that if we put $x = 2$ in the polynomial equation, we will obtain 12. This is done as follows.

$x^3 + 2x^2 + mx + n$

$(2)^3 + 2(2)^2 + m(2) + n = 0$

$8 + 8 + 2m + n = 12$

$2m + n = 12 - 8 - 8$

$2m + n = -4$ ....Equation (2)

From equation (1), $n = m - 1$ ....Equation (3)

Substitute $m - 1$ for n in equation (2)

$2m + n = -4$ ....Equation (2)

$2m + (m - 1) = -4$

$2m + m - 1 = -4$

$3m = -4 + 1$

$3m = -3$

$m = \dfrac{-3}{3}$

$m = -1$

Substitute −1 for m in equation (3)

$n = m - 1$ ....Equation (3)

$= -1 - 1$

$n = -2$

Hence, m = −1 and n = −2

b. The polynomial is $x^3 + 2x^2 + mx + n$. When we substitute the values of m and n the polynomial becomes:

$x^3 + 2x^2 - x - 2$ (Since m = −1 and n = −2)

From the question, (x + 1) is a factor of the polynomial. Hence, let us divide the polynomial by (x + 1) in order to get the other factors. The working is as shown below.

$$\begin{array}{r}
x^2 + x - 2 \\
x+1{\overline{\smash{\big)}\,x^3 + 2x^2 - x - 2}} \\
\underline{x^3 + x^2} \\
x^2 - x \\
\underline{x^2 + x} \\
-2x - 2 \\
\underline{-2x - 2} \\
- -
\end{array}$$

Hence the quadratic factor is $x^2 + x - 2$. We now factorize it as follows:

$x^2 + x - 2 = (x + 2)(x - 1)$

Let us now write the factorized polynomial as follows:

$x^3 + 2x^2 - x - 2 = (x + 1)(x + 2)(x - 1)$

c. Set the factorized polynomial to be equal to zero. This gives:

$x^3 + 2x^2 - x - 2 = 0$

$(x + 1)(x + 2)(x - 1) = 0$

Hence, x = −1, −2 or 1   (When each of the bracket above is equated to zero and solved)

Therefore the zeros of the polynomial are −1, −2 and 1.

10. If $2x^2 - 7x - 4$ is a factor of the polynomial $f(x) = 2x^4 - 5x^3 - 15x^2 + px + q$, where p and q are constants:

a. find the values of p and q

b. factorize f(x) completely.

Solution

a. Let us factorize $2x^2 - 7x - 4$ as follows:

$2x^2 - 7x - 4 = 2x^2 - 8x + x - 4$
$= 2x(x - 4) + 1(x - 4)$
$= (x - 4)(2x + 1)$

Hence $2x^2 - 7x - 4 = (x - 4)(2x + 1)$

Therefore $(x - 4)$ and $(2x + 1)$ are factors of $f(x)$

From these two factors, $x = 4$ and $x = -\frac{1}{2}$

Hence, $f(4) = 0$ and $f(-\frac{1}{2}) = 0$

$f(x) = 2x^4 - 5x^3 - 15x^2 + px + q$
$f(4) = 2(4)^4 - 5(4)^3 - 15(4)^2 + p(4) + q$

Since $f(4) = 0$, then it follows that:

$2(4)^4 - 5(4)^3 - 15(4)^2 + p(4) + q = 0$
$2(256) - 5(64) - 15(16) + 4p + q = 0$
$512 - 320 - 240 + 4p + q = 0$
$-48 + 4p + q = 0$
$4p + q = 48$ ....Equation (1)

Also, $f(x) = 2x^4 - 5x^3 - 15x^2 + px + q$

$f(-\frac{1}{2}) = 2(-\frac{1}{2})^4 - 5(-\frac{1}{2})^3 - 15(-\frac{1}{2})^2 + p(-\frac{1}{2}) + q$

Since $f(-\frac{1}{2}) = 0$, then it follows that:

$2(-\frac{1}{2})^4 - 5(-\frac{1}{2})^3 - 15(-\frac{1}{2})^2 + p(-\frac{1}{2}) + q = 0$

$2(\frac{1}{16}) - 5(-\frac{1}{8}) - 15(\frac{1}{4}) - \frac{p}{2} + q = 0$

$\frac{1}{8} + \frac{5}{8} - \frac{15}{4} - \frac{p}{2} + q = 0$

$\frac{1 + 5 - 30 - 4p + 8q}{8} = 0$

$-24 - 4p + 8q = 8(0)$
$-24 = 4p - 8q$
$4p - 8q = -24$
$p - 2q = -6$ (After dividing each term by 4)
$p - 2q = -6$ ....Equation (2)

From equation (2), $p = 2q - 6$ ....Equation (3)

Substitute $2q - 6$ for p in equation 1.

$4p + q = 48$ ....Equation (1)
$4(2q - 6) + q = 48$
$8q - 24 + q = 48$

$9q = 48 + 24$

$9q = 72$

$q = \dfrac{72}{9}$

$q = 8$

Substitute 8 for q in equation (3)

$p = 2q - 6$ ....Equation (3)

$= 2(8) - 6$

$= 16 - 6$

$p = 10$

Hence, p = 10 and q = 8

b. Substitute 10 for p and 8 for q in order to obtain the polynomial as follows:

$f(x) = 2x^4 - 5x^3 - 15x^2 + 10x + 8$

Since $2x^2 - 7x - 4$ is a factor of the polynomial, let us divide f(x) by $2x^2 - 7x - 4$ as shown below.

$$\begin{array}{r}
x^2 + x - 2 \\
2x^2 - 7x - 4 {\overline{\smash{\big)}\,2x^4 - 5x^3 - 15x^2 - 10x + 8}} \\
\underline{2x^4 - 7x^3 - 4x^2 } \\
2x^3 - 11x^2 + 10x \\
\underline{2x^3 - 7x^2 - 4x } \\
-4x^2 + 14x + 8 \\
\underline{-4x^2 + 14x + 8} \\
- \; - \; - \; - \; - \; -
\end{array}$$

Hence the other quadratic factor is $x^2 + x - 2$

We now factorize this as follows:

$x^2 + x - 2 = (x + 2)(x - 1)$

Recall that: $2x^2 - 7x - 4 = (x - 4)(2x + 1)$ (From our first step in (a) above)

Therefore the polynomial is completely factorized by using all our linear factors as follows:

$2x^4 - 5x^3 - 15x^2 + 10x + 8 = (x + 2)(x - 1)(x - 4)(2x + 1)$

Exercise 9

1. If $A = 3x^3 + 5x^2 - 1$, $B = 2x^3 + 7$, and $C = 2x^3 - x^2 + 2x - 4$, find:

a. A + B

b. 2C − 2B

c. 3C − A

d. $2B + A - 2C$

e. $A + 2B + C$

2. If $f(x) = x^3 - 4x^2 + 3x - 10$, find:

a. $f(-3)$

b. $f(0)$

3. If $A = 5x^3 - 2x^2 + x$ and $B = 2x^3 + 3x^2 - 5x - 2$, find AB.

4. If $M = 2x^3 - 3x^2 + 6x + 5$ and $N = 3x^3 - x^2 + 2x - 4$, find MN.

5. Given that $f(x) = x^3 - 5$ and $g(x) = 2x^3 - 3x^2 + 6x - 9$, find $f(x) \cdot g(x)$

6. Divide $3x^2 + 2x - 8$ by $3x - 4$

7. Divide $2x^3 + x^2 - 7x + 24$ by $x + 3$

8. Evaluate $\dfrac{6x^3 + 3x^2 - 10x - 5}{2x + 1}$

9. Divide $4x^3 - 12x^2y + 11xy^2 + 15y^3$ by $2x - 5y$

10. Simplify: $\dfrac{5x^3 + 8x^2y - 5xy^2 - 2y^3}{x + 2y}$

11. Find the quotient and the remainder when $2x^3 - 7x^2 + 12x - 9$ is divided by $x - 5$

12. Find the quotient and the remainder when $5x^3 + 2x^2 - 11x + 1$ is divided by $x + 2$

13. Find the zeros of the following polynomial functions:

a. $f(x) = 2x^2 - 19x + 30$

b. $f(x) = 4x^2 + 4x - 3$

14. Factorize $x^3 - 4x^2 + x + 6$ and use it to solve the cubic equation $x^3 - 4x^2 + x + 6 = 0$

15. Solve the equation $2x^4 + x^3 - 20x^2 - 13x + 30 = 0$

16. Find the remainder when $2x^3 - 9x^2 + 6x + 3$ is divided by $x - 4$.

17. Find the remainder when $5x^2 - 2x - 7$ is divided by $2x + 3$

18. If $(x + 1)$ and $(x - 2)$ are factors of $4x^4 + 4x^3 - 13x^2 - 19x - 6$, find the other two factors.

19. If $(2x - 1)$ is a factor of the polynomial $f(x) = 12x^3 + kx$, where k is a constant, find the zeros of $f(x)$.

20. Given that $2x + 1$ is a factor of the polynomial $6x^3 + 3x^2 - 10x - 5$, find the quadratic factor.

21. The remainder when the polynomial $f(x) = ax^3 + bx^2 + 5x - 9$ is divided by $x - 1$ is -10, and when it is divided by $x - 3$ the remainder is -12. Determine the values of a and b.

22. The polynomial $x^3 - 2x^2 + mx + n$ is divisible by $x - 1$. It leaves a remainder of -8 when divided by $x - 2$.

a. Find m and n

b. Find the quadratic factor

c. Find the zeros of the polynomial

23. If $2x^2 - x - 6$ is a factor of the polynomial $f(x) = 2x^4 + x^3 - 20x^2 + mx + n$, where m and n are constants. Find the values of m and n

CHAPTER 10
PARTIAL FRACTION

Consider the addition of the algebraic fractions below.

$$\frac{3}{x-2} + \frac{2}{x+5} = \frac{3(x+5)+2(x-2)}{(x-2)(x+5)}$$ [Note that the LCM is $(x-2)(x+5)$]

$$= \frac{3x+15+2x-4}{(x-2)(x+5)}$$

$$= \frac{5x+11}{(x-2)(x+5)}$$

Hence, $\frac{3}{x-2}$ and $\frac{2}{x+5}$ added up to give $\frac{5x+11}{(x-2)(x+5)}$

We say that $\frac{3}{x-2}$ and $\frac{2}{x+5}$ are the partial fractions of $\frac{5x+11}{(x-2)(x+5)}$. The result obtained i.e. $\frac{5x+11}{(x-2)(x+5)}$ is called the compound fraction. Since we know how to move from partial fractions to compound fraction, it is also important that we know how to move from compound fraction to partial fractions. This is what we want to deal with here.

Resolving Algebraic Fractions into Partial Fractions

The steps below are applied in resolving algebraic fraction into partial fractions:

1. The polynomial must be factorized or already be expressed in factors.
2. The highest power of the variable in the numerator must be at least one less than that of the denominator.
3. When the highest power of the variable in the numerator is equal to, or higher than the highest power of the variable in the denominator, then the numerator should be divided by the denominator before resolving the remainder (expressed as fraction) into partial fractions.

Types of Partial Fraction

1. **LINEAR FACTORS:** Algebraic fractions whose denominators contain linear factors are resolved into partial fractions as shown below.

$$\frac{f(x)}{(x+a)(x+b)} = \frac{A}{x+a} + \frac{B}{x+b}$$

2. **REPEATED LINEAR FACTORS:** Fractions with repeated linear denominators are resolved as follows:

$$\frac{f(x)}{(x+a)^2} = \frac{A}{x+a} + \frac{B}{(x+a)^2}$$

And:

$$\frac{f(x)}{(x+a)^3} = \frac{A}{x+a} + \frac{B}{(x+a)^2} + \frac{C}{(x+a)^3}$$

3. **QUADRATIC FACTORS:** When a denominator of an algebraic fraction contains a quadratic expression that cannot be factorized into linear factors, then it is resolved into partial fractions as follows:

$$\frac{f(x)}{(ax^2+bx+c)(x+d)} = \frac{Ax+B}{(ax^2+bx+c)} + \frac{C}{x+d}$$

Note that in all three cases above, the highest power of the variable in f(x) must be less than the highest power of the variable in the denominator. Otherwise we must divide first before we proceed.

Examples

1. Express $\frac{x+21}{(x+3)(x+1)}$ in partial fractions

<u>Solution</u>

METHOD 1

Let $\frac{x+21}{(x+3)(x+1)} = \frac{A}{x+3} + \frac{B}{x+1}$

The LCM or LCD (lowest common denominator) of the right hand side is $(x+3)(x+1)$. Hence we carry out the addition of the right hand side as follows:

$$\frac{x+21}{(x+3)(x+1)} = \frac{A(x+1)+B(x+3)}{(x+3)(x+1)}$$

Hence, since the denominators are the same on both sides of the equation, then the numerators should also be the same or identical. Hence for the numerators, we write:

$x + 21 \equiv A(x + 1) + B(x + 3)$

This is an identity. It means that both sides are equal for all values of x, hence the use of the equivalent symbol. In order to find A and B, we take a value of x that will make the term in any of the bracket to be zero. In order to find the number that will make A to be zero, we simply equate the bracket attached to A to be equal to zero and solve for x. For example, $x + 1 = 0$, which will give $x = -1$ when we solve for x. Hence if we put $x = -1$, A will be eliminated so that we can obtain B. This gives:

$x + 21 \equiv A(x + 1) + B(x + 3)$

$-1 + 21 = A(-1 + 1) + B(-1 + 3)$

$20 = A(0) + 2B$

$20 = 2B$ (Since A x 0 = 0)

$B = \frac{20}{2}$

$B = 10$

If we put $x = -3$ (from $x + 3 = 0$, then $x = -3$), B will be eliminated, so that we can obtain A. This gives:

$x + 21 \equiv A(x + 1) + B(x + 3)$

$-3 + 21 = A(-3 + 1) + B(-3 + 3)$

$18 = A(-2) + B(0)$

$18 = -2A$ (Since B x 0 = 0)

$A = \dfrac{18}{-2}$

$A = -9$

Hence, A = –9 and B = 10

Therefore, $\dfrac{x+21}{(x+3)(x+1)} = \dfrac{-9}{x+3} + \dfrac{10}{x+1}$

Or, $\dfrac{x+21}{(x+3)(x+1)} = \dfrac{10}{x+1} - \dfrac{9}{x+3}$ (When we rearrange the answer above)

METHOD 2

In this method, we follow the same procedure as in method 1 until we get to the point written below:

$x + 21 \equiv A(x + 1) + B(x + 3)$

Now we expand the brackets above, then simplify and collect like terms as carried out below.

$x + 21 \equiv Ax + A + Bx + 3B$

$x + 21 \equiv Ax + Bx + A + 3B$ (By collecting like terms together)

$x + 21 \equiv (A + B)x + A + 3B$ (After factorizing terms having like variables)

By comparing coefficients of like terms on both sides of the identity, we have:

A + B = 1

On the left hand side of the identity above, we have $1x$, while on the right hand side we have $(A + B)x$. Hence the coefficients of x are 1 and A + B. Therefore, A + B = 1.

Similarly, A + 3B = 21

The constant term on the left hand side is 21, while the constant terms on the right hand side are A and 3B. Hence we equate them to obtain A + 3B = 21.

Therefore our two equations above are:

$\qquad\qquad$ A + B = 1 ……………Equation (1)

$\qquad\qquad$ A + 3B = 21 ……………..Equation (2)

Equation (2) – equation (1):   2B = 20

$B = \dfrac{20}{2}$

B = 10

Substitute 10 for B in equation (1).

$\quad$ A + B = 1 ……………Equation (1)

$\quad$ A + 10 = 1

$\quad$ A = 1 – 10

$\quad$ A = –9

Therefore, A = –9 and B = 10

Hence, $\dfrac{x+21}{(x+3)(x+1)} = \dfrac{-9}{x+3} + \dfrac{10}{x+1}$

Or, $\dfrac{x+21}{(x+3)(x+1)} = \dfrac{10}{x+1} - \dfrac{9}{x+3}$ (As obtained in method 1 above)

It is obvious that method 1 is more direct or shorter than method 2, since method 2 leads to a simultaneous equation. Hence, in most of our examples we will be using method 1. However in some cases where only method 1 cannot be used, we will apply the two methods or only method 2, as the case may be.

2. Resolve $\dfrac{4x-16}{x^2-2x-3}$ into partial fractions.

Solution

The denominator can be factorized into linear factors as follows:

$x^2 - 2x - 3 = (x-3)(x+1)$

Hence the fraction is now written as follows:

$\dfrac{4x-16}{x^2-2x-3} = \dfrac{4x-16}{(x-3)(x+1)}$

We now resolve the factorized one into partial fractions as follows:

Let $\dfrac{4x-16}{(x-3)(x+1)} = \dfrac{A}{x-3} + \dfrac{B}{x+1}$

$\dfrac{4x-16}{(x-3)(x+1)} = \dfrac{A(x+1)+B(x-3)}{(x-3)(x+1)}$

Since the denominators are the same, then the numerators are identical and can be brought out to give:

$4x - 16 \equiv A(x+1) + B(x-3)$

Now substitute $x = -1$ into the identity in order to obtain B as follows:

$4x - 16 \equiv A(x+1) + B(x-3)$
$4(-1) - 16 = A(-1+1) + B(-1-3)$
$-4 - 16 = A(0) - 4B$
$-20 = -4B$ (Since A x 0 = 0)
$B = \dfrac{-20}{-4}$
$B = 5$

Put $x = 3$ in order to obtain A as follows:

$4x - 16 \equiv A(x+1) + B(x-3)$
$4(3) - 16 = A(3+1) + B(3-3)$
$12 - 16 = A(4) + B(0)$
$-4 = 4A$ (Since B x 0 = 0)
$A = \dfrac{-4}{4}$
$A = -1$

Hence, A = −1 and B = 5

Therefore, $\dfrac{4x-16}{x^2-2x-3} = -\dfrac{1}{x-3} + \dfrac{5}{x+1}$

$$= \frac{5}{x+1} - \frac{1}{x-3} \quad \text{(When we rearrange the answer above)}$$

3. Express $\frac{15}{x^2-9}$ in partial fractions.

Solution

The denominator is a difference of two squares. Recall that a difference of two squares such as $a^2 - b^2$ can be factorized as follows:

$a^2 - b^2 = (a+b)(a-b)$

Similarly, the denominator above can be expressed as a difference of two squares and factorized as follows:

$x^2 - b^2 = x^2 - 3^2 = (x+3)(x-3)$ (Note that 9 has been expressed as 3^2)

Hence, $\frac{15}{(x+3)(x-3)} = \frac{A}{x+3} + \frac{B}{x-3}$

$\frac{15}{(x+3)(x-3)} = \frac{A(x-3) + B(x+3)}{(x+3)(x-3)}$

$15 \equiv A(x-3) + B(x+3)$

Substitute $x = 3$ into the identity to obtain B as follows:

$15 = A(3-3) + B(3+3)$ (Note that the right hand side remains 15 since there is no x there)

$15 = 0 + 6B$

$15 = 6B$

$B = \frac{15}{6}$

$B = \frac{5}{2}$ (In its lowest term)

Put $x = -3$ in order to eliminate B and obtain A as follows:

$15 \equiv A(x-3) + B(x+3)$

$15 = A(-3-3) + B(-3+3)$

$15 = -6A + 0$

$15 = -6A$

$A = \frac{15}{-6}$

$A = -\frac{5}{2}$

Hence, $A = -\frac{5}{2}$ and $B = \frac{5}{2}$

Therefore, $\frac{15}{x^2-9} = \frac{-\frac{5}{2}}{x+3} + \frac{\frac{5}{2}}{x-3}$

$= -\frac{5}{2(x+3)} + \frac{5}{2(x-3)}$

$= \frac{5}{2(x-3)} - \frac{5}{2(x+3)}$

4. Resolve $\dfrac{2x-9}{x(x-3)}$ into partial fractions.

Solution

$$\dfrac{2x-9}{x(x-3)} = \dfrac{A}{x} + \dfrac{B}{x-3}$$

$$\dfrac{2x-9}{x(x-3)} = \dfrac{A(x-3)+B(x)}{x(x-3)}$$

Equating the numerators gives:

$$2x - 9 \equiv A(x-3) + B(x)$$

Substitute $x = 3$ into the identity to obtain B as follows:

2(3) − 9 = A(3 − 3) + B(3)

6 − 9 = 0 + 3B

−3 = 3B

$B = \dfrac{-3}{3}$

B = −1

Put $x = 0$ in order to eliminate B and obtain A as follows:

$2x - 9 \equiv A(x-3) + B(x)$

2(0) − 9 = A(0 − 3) + B(0)

0 − 9 = −3A + 0

−9 = −3A

$A = \dfrac{-9}{-3}$

A = 3

Hence, A = 3 and B = −1

Therefore, $\dfrac{2x-9}{x(x-3)} = \dfrac{3}{x} - \dfrac{1}{x-3}$

5. Resolve $\dfrac{2x+12}{(x+5)^2}$ into partial fractions.

Solution

The denominator is a repeated linear factor. Hence we resolve into partial factions as follows:

$$\dfrac{2x+12}{(x+5)^2} = \dfrac{A}{x+5} + \dfrac{B}{(x+5)^2}$$

$$\dfrac{2x+12}{(x+5)^2} = \dfrac{A(x+5)+B(1)}{(x+5)^2}$$ [Note that $(x+5)^2 = (x+5)(x+5)$]

Equating the numerators gives:

$$2x + 12 \equiv A(x+5) + B$$

Substitute −5 for x in order to solve for B

2(−5) + 12 = A(−5 + 5) + B

−10 + 12 = A(0) + B

2 = B
B = 2

Since we already know the value of B, we can substitute any value for x into the identity above and obtain the value of A. Note that an identity is true for all values of the variable. Hence let us substitute 1 for x (any value of x can be used) into the identity while also substituting 2 for B as we have already obtained above. This will give us A as follows:

$2x + 12 \equiv A(x + 5) + B$
$2(1) + 12 = A(1 + 5) + 2$
$2 + 12 = 6A + 2$
$14 - 2 = 6A$
$12 = 6A$
$A = \dfrac{12}{6}$
$A = 2$

Hence, A = 2 and B = 2

Therefore, $\dfrac{2x + 12}{(x + 5)^2} = \dfrac{2}{x + 5} + \dfrac{2}{(x + 5)^2}$

6. Express $\dfrac{3x-5}{x^2(x-2)}$ in partial fractions.

<u>Solution</u>

x^2 in the denominator is like a repeated linear factor. With that in mind, we express in partial fraction as follows:

$\dfrac{3x-5}{x^2(x-2)} = \dfrac{A}{x} + \dfrac{B}{x^2} + \dfrac{C}{x-2}$

$\dfrac{3x-5}{x^2(x-2)} = \dfrac{A[x(x-2)] + B(x-2) + C(x^2)}{x^2(x-2)}$

Equating the numerators gives:

$3x - 5 \equiv A[x(x-2)] + B(x-2) + Cx^2$

Observing the right hand side of the identity shows that if we substitute $x = 0$, we will obtain the value of B as follows:

$3x - 5 \equiv A[x(x-2)] + B(x-2) + Cx^2$
$3(0) - 5 = A[0(0-2)] + B(0-2) + C(0^2)$
$0 - 5 = A[0(-2)] + B(-2) + 0$
$-5 = 0 - 2B + 0$
$-5 = -2B$
$B = \dfrac{-5}{-2}$
$B = \dfrac{5}{2}$

Similarly, substituting 2 for x in the identity above will give us the value of C as follows:

$$3x - 5 \equiv A[x(x - 2)] + B(x - 2) + Cx^2$$
$$3(2) - 5 = A[2(2 - 2)] + B(2 - 2) + C(2^2)$$
$$6 - 5 = A[2(0)] + B(0) + 4C$$
$$1 = 0 + 0 + 4C$$
$$1 = 4C$$
$$C = \frac{1}{4}$$

Now, since we already know the values of B and C, we can substitute any value of x (apart from 0 and 2 used above) into the identity in order to obtain the last letter which is A. Note that this can be done when finding the last letter in any identity. Hence, let us put $x = 1$ into the identity as follows:

$$3x - 5 \equiv A[x(x - 2)] + B(x - 2) + Cx^2$$
$$3(1) - 5 = A[1(1 - 2)] + \frac{5}{2}(1 - 2) + \frac{1}{4}(1^2)$$
$$3 - 5 = A[1(-1)] + \frac{5}{2}(-1) + \frac{1}{4}(1)$$
$$-2 = -A - \frac{5}{2} + \frac{1}{4}$$
$$A = 2 - \frac{5}{2} + \frac{1}{4}$$
$$A = \frac{8 - 10 + 1}{4}$$
$$A = -\frac{1}{4}$$

Hence, $A = -\frac{1}{4}$, $B = \frac{5}{2}$ and $C = \frac{1}{4}$

Therefore, $\dfrac{3x - 5}{x^2(x-2)} = \dfrac{-\frac{1}{4}}{x} + \dfrac{\frac{5}{2}}{x^2} + \dfrac{\frac{1}{4}}{x - 2}$

$$= -\frac{1}{4x} + \frac{5}{2x^2} + \frac{1}{4(x - 2)}$$

This can also be rearranged to avoid starting with a negative value as follows:

$$\frac{3x - 5}{x^2(x-2)} = \frac{5}{2x^2} - \frac{1}{4x} + \frac{1}{4(x - 2)}$$

7. Resolve $\dfrac{x + 6}{(x + 2)(x - 3)^2}$ into partial fractions.

<u>Solutions</u>

The denominator contains repeated factors. Hence we resolve as follows:

$$\frac{x + 6}{(x + 2)(x - 3)^2} = \frac{A}{x + 2} + \frac{B}{x - 3} + \frac{C}{(x - 3)^2}$$

$$\frac{x + 6}{(x + 2)(x - 3)^2} = \frac{A(x-3)^2 + B(x + 2)(x - 3) + C(x+2)}{(x + 2)(x - 3)^2}$$

Equating the numerators gives:

$$x + 6 \equiv A(x - 3)^2 + B(x + 2)(x - 3) + C(x + 2)$$

Substitute 3 for x in order to obtain C as follows:

$3 + 6 = A(3 - 3)^2 + B(3 + 2)(3 - 3) + C(3 + 2)$

$9 = A(0) + B(5)(0) + C(5)$

$9 = 0 + 0 + 5C$

$9 = 5C$

$C = \dfrac{9}{5}$

Substitute -2 for x in the identity above in order to find A. This gives:

$x + 6 \equiv A(x - 3)^2 + B(x + 2)(x - 3) + C(x + 2)$

$-2 + 6 = A(-2 - 3)^2 + B(-2 + 2)(-2 - 3) + C(-2 + 2)$

$4 = A(-5)^2 + B(0)(-5) + C(0)$

$4 = A(25) + 0 + 0$

$4 = 25A$

$A = \dfrac{4}{25}$

Since we now know the values of A and C, we can substitute any value for x in the identity above in order to get B. However, I want us to use the method of comparing coefficient to find B. In order to use this method, we have to expand the brackets on the right hand side of the identity. Let us expand and then compare coefficients as follows:

$x + 6 \equiv A(x - 3)^2 + B(x + 2)(x - 3) + C(x + 2)$

$x + 6 \equiv A(x - 3)(x - 3) + B(x + 2)(x - 3) + C(x + 2)$

$x + 6 \equiv A(x^2 - 6x + 9) + B(x^2 - x - 6) + Cx + 2C$

$x + 6 \equiv Ax^2 - 6Ax + 9A + Bx^2 - Bx - 6B + Cx + 2C$

$x + 6 \equiv (A + B)x^2 - (6A + B - C)x + (9A - 6B + 2C)$

By comparing the coefficient of x^2 on both sides of the identity shows that:

$A + B = 0$

Since there is no term in x^2 on the left hand side of the identity, it means that the coefficient of x^2 on the left hand side is zero. The coefficient of x^2 on the right hand side is A + B. Hence equating the two coefficients gives A + B = 0, as written above.

We already know the value of A from our answer above, hence we calculate B as follows:

$A + B = 0$

$\dfrac{4}{25} + B = 0$

$B = -\dfrac{4}{25}$

Hence, $A = \dfrac{4}{25}$, $B = -\dfrac{4}{25}$ and $C = \dfrac{9}{5}$

Therefore, $\dfrac{x + 6}{(x + 2)(x - 3)^2} = \dfrac{4}{25(x + 2)} - \dfrac{4}{25(x - 3)} + \dfrac{9}{5(x - 3)^2}$

Or, $\dfrac{x + 6}{(x + 2)(x - 3)^2} = \dfrac{1}{5}\left[\dfrac{4}{5(x + 2)} - \dfrac{4}{5(x - 3)} + \dfrac{9}{(x - 3)^2}\right]$

8. Resolve $\dfrac{2x^2-3x-4}{x(x^2-2)}$ into partial fractions.

Solution

Note that the denominator contains a quadratic factor which does not factorize. This means that when the quadratic factor is resolved to partial fraction, its numerator will contain x and two constant terms. Hence we resolve into partial fractions as follows:

$$\dfrac{2x^2-3x-4}{x(x^2-2)} = \dfrac{A}{x} + \dfrac{Bx+C}{x^2-2}$$

$$\dfrac{2x^2-3x-4}{x(x^2-2)} = \dfrac{A(x^2-2)+(Bx+C)(x)}{x(x^2-2)}$$

Equating the numerators gives:

$$2x^2 - 3x - 4 \equiv A(x^2 - 2) + (Bx + C)(x)$$

Substitute 0 for x in the identity above. This gives:

$2(0)^2 - 3(0) - 4 = A((0)^2 - 2) + [B(0) + C](0)$

$0 - 0 - 4 = A(0 - 2) + 0$

$-4 = -2A$

$A = \dfrac{-4}{-2}$

$A = 2$

In the identity above, we can no longer substitute any value of x in order to obtain B or C. Hence we have to use the method of comparing coefficients to find B and C. In order to use this method, we expand the bracket in the identity above. This gives:

$2x^2 - 3x - 4 \equiv A(x^2 - 2) + (Bx + C)(x)$

$\equiv Ax^2 - 2A + Bx^2 + Cx$

$\equiv Ax^2 + Bx^2 + Cx - 2A$

$2x^2 - 3x - 4 \equiv (A + B)x^2 + Cx - 2A$ [Note that $Ax^2 + Bx^2$ has been factorized to give $(A + B)x^2$]

Comparing the coefficients of x^2 on both sides of the identity shows that A + B = 2. This is because on the left hand side we have $2x^2$ and on the right hand side we have $(A + B)x^2$. Hence $2x^2 = (A + B)x^2$, which shows that A + B = 2.

Let us solve this equation as follows:

A + B = 2

2 + B = 2   (Since A = 2 as obtained above)

B = 2 – 2

B = 0

Similarly, comparing the coefficients of x in the identity above, shows that:

C = –3   (Since $-3x = Cx$)

Hence, A = 2, B = 0 and C = –3

Therefore, $\dfrac{2x^2-3x-4}{x(x^2-2)} = \dfrac{2}{x} + \dfrac{0x-3}{x^2-2}$

This gives: $\dfrac{2x^2 - 3x - 4}{x(x^2 - 2)} = \dfrac{2}{x} - \dfrac{3}{x^2 - 2}$

9. Resolve $\dfrac{2x^2 - 1}{(x^2 - 1)(x - 2)}$ into partial fractions

Solution

$$\dfrac{2x^2 - 1}{(x^2 - 1)(x - 2)} = \dfrac{Ax + B}{(x^2 - 1)} + \dfrac{C}{(x - 2)}$$

$$\dfrac{2x^2 - 1}{(x^2 - 1)(x - 2)} = \dfrac{(Ax + B)(x - 2) + C(x^2 - 1)}{(x^2 - 1)(x - 2)}$$

Equating the numerators gives:

$2x^2 - 1 \equiv (Ax + B)(x - 2) + C(x^2 - 1)$

Substitute 2 for x in order to get C. This gives:

$2(2)^2 - 1 = (A(2) + B)(2 - 2) + C(2^2 - 1)$

$2(4) - 1 = (2A + B)(0) + C(4 - 1)$

$8 - 1 = 0 + 3C$

$7 = 3C$

$C = \dfrac{7}{3}$

Now let us expand the bracket in the identity above.

$2x^2 - 1 \equiv (Ax + B)(x - 2) + C(x^2 - 1)$
$\equiv Ax^2 - 2Ax + Bx - 2B + Cx^2 - C$
$\equiv Ax^2 + Cx^2 - 2Ax + Bx - 2B - C$

$2x^2 - 1 \equiv (A + C)x^2 + (-2A + B)x - (2B + C)$ (After factorization)

Comparing coefficient of terms on both sides of the identity shows that:

A + C = 2 ……………..Equation (1)

−2A + B = 0 ……………..Equation (2)

Note that there is no term in x on the left hand side, hence the coefficient of x on the left hand side is zero. This is why equation (2) is equated to zero as shown above.

−(2B + C) = −1   (This is obtained by comparing the coefficients of the constant terms)

Or, 2B + C = 1 ……………..Equation (3)   (After dividing both sides by −1)

Hence from equation (1), we have:

A + C = 2 ……………..Equation (1)

$A + \dfrac{7}{3} = 2$ (Since $C = \dfrac{7}{3}$)

$A = 2 - \dfrac{7}{3}$

$= \dfrac{6 - 7}{3}$

$A = -\frac{1}{3}$

From equation (2) we have:

$-2A + B = 0$ ....Equation (2)

$-2(-\frac{1}{3}) + B = 0$

$\frac{2}{3} + B = 0$

$B = -\frac{2}{3}$

Hence, $A = -\frac{1}{3}$, $B = -\frac{2}{3}$ and $C = \frac{7}{3}$

Therefore, $\dfrac{2x^2 - 1}{(x^2 - 1)(x - 2)} = \dfrac{-\frac{1}{3}x - \frac{2}{3}}{(x^2 - 1)} + \dfrac{\frac{7}{3}}{(x - 2)}$

$= \dfrac{-\frac{1}{3}(x + 2)}{(x^2 - 1)} + \dfrac{7}{3(x - 2)}$

$\dfrac{2x^2 - 1}{(x^2 - 1)(x - 2)} = \dfrac{7}{3(x - 2)} - \dfrac{(x + 2)}{3(x^2 - 1)}$

Note that the method of comparing coefficients is often used when we have a quadratic factor in the denominator.

10. Resolve $\dfrac{6x^2 - 24x - 3}{(2x - 1)(x^2 + 5x + 4)}$ into partial fractions

<u>Solution</u>

A careful look at the quadratic part of the denominator shows that it can be factorized into linear factors as follows:

$x^2 + 5x + 4 = (x + 4)(x + 1)$

Hence the fraction in the question above can be rewritten as:

$\dfrac{6x^2 - 24x - 3}{(2x - 1)(x + 4)(x + 1)}$

We now resolve into partial fractions as follows:

$\dfrac{6x^2 - 24x - 3}{(2x - 1)(x + 4)(x + 1)} = \dfrac{A}{2x - 1} + \dfrac{B}{x + 4} + \dfrac{C}{x + 1}$

$\dfrac{6x^2 - 24x - 3}{(2x - 1)(x + 4)(x + 1)} = \dfrac{A(x + 4)(x + 1) + B(2x - 1)(x + 1) + C(2x - 1)(x + 4)}{(2x - 1)(x + 4)(x + 1)}$

Equating the numerators gives:

$6x^2 - 24x - 3 \equiv A(x + 4)(x + 1) + B(2x - 1)(x + 1) + C(2x - 1)(x + 4)$

Substitute $x = -4$ in order to obtain B. This gives:

$6(-4)^2 - 24(-4) - 3 = A(-4 + 4)(-4 + 1) + B(2(-4) - 1)(-4 + 1) + C(2(-4) - 1)(-4 + 4)$

$6(16) + 96 - 3 = A(0)(-3) + B(-8 - 1)(-3) + C(-8 - 1)(0)$

$96 + 96 - 3 = 0 + 27B + 0$

189 = 27B
$$B = \frac{189}{27}$$
B = 7

Substitute –1 for x in the identity above. This will give C as follows:

$6x^2 - 24x - 3 \equiv A(x + 4)(x + 1) + B(2x - 1)(x + 1) + C(2x - 1)(x + 4)$

$6(-1)^2 - 24(-1) - 3 = A(-1 + 4)(-1 + 1) + B(2(-1) - 1)(-1 + 1) + C(2(-1) - 1)(-1 + 4)$

$6 + 24 - 3 = A(3)(0) + B(-2 - 1)(0) + C(-2 - 1)(3)$

$27 = 0 + 0 + C(-3)(3)$

$27 = -9C$

$$C = \frac{27}{-9}$$

C = –3

Since we already know B and C we can substitute any value for x in the identity and obtain A. Let us substitute 1 for x in order to obtain A. This gives:

$6x^2 - 24x - 3 \equiv A(x + 4)(x + 1) + B(2x - 1)(x + 1) + C(2x - 1)(x + 4)$

$6(1)^2 - 24(1) - 3 = A(1 + 4)(1 + 1) + B(2(1) - 1)(1 + 1) + C(2(1) - 1)(1 + 4)$

$6 - 24 - 3 = A(5)(2) + B(2 - 1)(2) + C(2 - 1)(5)$

$-21 = 10A + 2B + 5C$

$-21 = 10A + 2(7) + 5(-3)$ (Note that B = 7 and C = –3)

$-21 = 10A + 14 - 15$

$-21 = 10A - 1$

$-21 + 1 = 10A$

$-20 = 10A$

$$A = \frac{-20}{10}$$

A = –2

Hence A = –2, B = 7 and C = –3

Therefore, $\dfrac{6x^2 - 24x - 3}{(2x - 1)(x^2 + 5x + 4)} = -\dfrac{2}{2x - 1} + \dfrac{7}{x + 4} - \dfrac{3}{x + 1}$

Or, $\dfrac{6x^2 - 24x - 3}{(2x - 1)(x^2 + 5x + 4)} = \dfrac{7}{x + 4} - \dfrac{2}{2x - 1} - \dfrac{3}{x + 1}$

11. Express $\dfrac{6x^2 + 19x - 11}{(x + 1)(x^2 + 5x - 2)}$ in partial fractions

<u>Solution</u>

The quadratic factor in the denominator cannot be factorized into linear factors. Therefore we resolve into partial fractions as follows:

$$\dfrac{6x^2 + 19x - 11}{(x + 1)(x^2 + 5x - 2)} = \dfrac{A}{x + 1} + \dfrac{Bx + C}{(x^2 + 5x - 2)}$$

$$\frac{6x^2+19x-11}{(x+1)(x^2+5x-2)} = \frac{A(x^2+5x-2)+(Bx+C)(x+1)}{(x+1)(x^2+5x-2)}$$

Equating the numerators gives:

$6x^2 + 19x - 11 \equiv A(x^2 + 5x - 2) + (Bx + C)(x + 1)$

Substitute –1 for x in order to obtain A as follows:

$6(-1)^2 + 19(-1) - 11 = A((-1)^2 + 5(-1) - 2) + (B(-1) + C)(-1 + 1)$

$6 - 19 - 11 = A(1 - 5 - 2) + (-B + C)(0)$

$-24 = -6A + 0$

$6A = 24$

$A = \dfrac{24}{6}$

$A = 4$

Since we have a quadratic factor in this question, we will have to apply the method of comparing coefficients. Hence, let us expand the brackets in the identity above. This gives:

$6x^2 + 19x - 11 \equiv A(x^2 + 5x - 2) + (Bx + C)(x + 1)$

$6x^2 + 19x - 11 \equiv Ax^2 + 5Ax - 2A + Bx^2 + Bx + Cx + C$

$\equiv Ax^2 + Bx^2 + 5Ax + Bx + Cx - 2A + C$

$6x^2 + 19x - 11 \equiv (A + B)x^2 + (5A + B + C)x - 2A + C$

Comparing the coefficients of like terms on both sides of the equation shows that:

A + B = 6 ……………….. Equation (1)

5A + B + C = 19 …………………Equation (2)

We have already calculated the value of A to be 4 as done above. Hence from equation (1) we have:

A + B = 6 ……………..Equation (1)

4 + B = 6   (Since A = 4)

B = 6 – 4

B = 2

From equation (2) we have:

5A + B + C = 19 ……………….Equation (2)

5(4) + 2 + C = 19

20 + 2 + C = 19

C = 19 – 22

C = –3

Hence, A = 4, B = 2 and C = –3

$$\frac{6x^2+19x-11}{(x+1)(x^2+5x-2)} = \frac{A}{x+1} + \frac{Bx+C}{(x^2+5x-2)}$$

Therefore, $\dfrac{6x^2+19x-11}{(x+1)(x^2+5x-2)} = \dfrac{4}{x+1} + \dfrac{2x-3}{(x^2+5x-2)}$

12. Resolve $\dfrac{2x^2-5x+1}{(x-2)^3}$ into partial fractions

Solution

This is a case of repeated linear factor. Hence we resolve into partial fractions as follows:

$$\dfrac{2x^2-5x+1}{(x-2)^3} = \dfrac{A}{(x-2)} + \dfrac{B}{(x-2)^2} + \dfrac{C}{(x-2)^3}$$

$$\dfrac{2x^2-5x+1}{(x-2)^3} = \dfrac{A(x-2)^2 + B(x-2) + C}{(x-2)^3}$$

Equating the numerators gives:

$2x^2 - 5x + 1 \equiv A(x-2)^2 + B(x-2) + C$

Substitute 2 for x in order to eliminate A and B and obtain the value of C. This gives:

$2(2)^2 - 5(2) + 1 = A(2-2)^2 + B(2-2) + C$

$2(4) - 10 + 1 = A(0) + B(0) + C$

$8 - 10 + 1 = 0 + 0 + C$

$-1 = C$

$C = -1$

In order to find A and B, we need to compare coefficients. Let us expand the brackets in the identity above as follows:

$2x^2 - 5x + 1 \equiv A(x-2)^2 + B(x-2) + C$

$\equiv A(x-2)(x-2) + B(x-2) + C$

$\equiv A(x^2 - 4x + 4) + Bx - 2B + C$

$\equiv Ax^2 - 4Ax + 4A + Bx - 2B + C$

$2x^2 - 5x + 1 \equiv Ax^2 + (-4A + B)x + 4A - 2B + C$

Comparing the coefficients of like terms on both sides of the equation shows that:

A = 2   (Since $2x^2 \equiv Ax^2$)

Similarly, $-4A + B = -5$ (From $-5x \equiv (-4A + B)x$)

$-4(2) + B = -5$ (Note that A = 2 as obtained above)

$-8 + B = -5$

$B = -5 + 8$

$B = 3$

Hence, A = 2, B = 3 and C = −1

Therefore, $\dfrac{2x^2-5x+1}{(x-2)^3} = \dfrac{2}{(x-2)} + \dfrac{3}{(x-2)^2} - \dfrac{1}{(x-2)^3}$

13. Resolve $\dfrac{5x^3+3x^2+2x-1}{x^2(x^2+1)}$ into partial fractions

Solution

Let $\dfrac{5x^3+3x^2+2x-1}{x^2(x^2+1)} = \dfrac{A}{x} + \dfrac{B}{x^2} + \dfrac{Cx+D}{x^2+1}$

$$\frac{5x^3+3x^2+2x-1}{x^2(x^2+1)} = \frac{A[x(x^2+1)]+B(x^2+1)+(Cx+D)x^2}{x^2(x^2+1)}$$

Equating the numerators gives:

$$5x^3 + 3x^2 + 2x - 1 \equiv A[x(x^2+1)] + B(x^2+1) + (Cx+D)(x^2)$$
$$\equiv A(x^3+x) + B(x^2+1) + Cx^3 + Dx^2$$
$$\equiv Ax^3 + Ax + Bx^2 + B + Cx^3 + Dx^2$$
$$\equiv Ax^3 + Cx^3 + Bx^2 + Dx^2 + Ax + B$$
$$5x^3 + 3x^2 + 2x - 1 \equiv (A+C)x^3 + (B+D)x^2 + Ax + B$$

Comparing the coefficients of terms on both sides of the identity, gives:

$A = 2$ (From coefficients of x)
$B = -1$ (The constant term on both sides of the identity)
$A + C = 5$ (From coefficients of x^3)
$2 + C = 5$ (Since $A = 2$)
$C = 5 - 2$
$C = 3$

Similarly, $B + D = 3$ (From coefficients of x^2)
$-1 + D = 3$
$D = 3 + 1$
$D = 4$

Hence, $A = 2$, $B = -1$, $C = 3$ and $D = 4$

Therefore, $\dfrac{5x^3+3x^2+2x-1}{x^2(x^2+1)} = \dfrac{2}{x} - \dfrac{1}{x^2} + \dfrac{3x+4}{x^2+1}$

14   Resolve $\dfrac{2x^2+19x+47}{x^2+9x+20}$ into partial fractions.

<u>Solutions</u>

From the question the highest power of x in the numerator and denominator are equal (i.e. x^2 in both cases). Hence we have to divide the numerator by the denominator in order to reduce the power of x in the numerator. Let us divide it as follows:

$$\begin{array}{r}2\\x^2+9x+20\overline{)2x^2+19x+47}\\2x^2+18x+40\\\hline x+7\end{array}$$

We cannot divide further, hence:

$$\dfrac{2x^2+19x+47}{x^2+9x+20} = 2 + \dfrac{x+7}{x^2+9x+20}$$

Note that 2 is the quotient and $x + 7$ is the remainder in the division carried out above.

Hence we now resolve $\dfrac{x+7}{x^2+9x+20}$ into partial fractions.

The denominator factorizes into linear factors as follows:
$$x^2 + 9x + 20 = (x + 4)(x + 5)$$
Hence the fraction above can be written as:
$$\frac{x+7}{(x+4)(x+5)}$$
Therefore, $\dfrac{x+7}{(x+4)(x+5)} = \dfrac{A}{(x+4)} + \dfrac{B}{(x+5)}$

$\dfrac{x+7}{(x+4)(x+5)} = \dfrac{A(x+5) + B(x+4)}{(x+4)(x+5)}$

Equating the numerators gives:
$$x + 7 \equiv A(x + 5) + B(x + 4)$$
Substitute $x = -5$ in order to get B. This gives:
$-5 + 7 = A(-5 + 5) + B(-5 + 4)$
$2 = A(0) + B(-1)$
$2 = 0 - B$
$B = -2$

Substitute $x = -4$ in order to find A. This gives:
$x + 7 \equiv A(x + 5) + B(x + 4)$
$-4 + 7 = A(-4 + 5) + B(-4 + 4)$
$3 = A(1) + B(0)$
$3 = A + 0$
$A = 3$

Hence A = 3 and B = −2

Therefore, $\dfrac{2x^2 + 19x + 47}{x^2 + 9x + 20} = 2 + \dfrac{x+7}{x^2 + 9x + 20}$ (Note: 2 was obtained from the division above)

$= 2 + \dfrac{3}{(x+4)} - \dfrac{2}{(x+5)}$

15. Resolve $\dfrac{x^3 + 2x^2 - x - 11}{x^2 - x - 2}$ into partial fractions

Solution

From the question, the highest power of x in the numerator is higher than the highest power of x in the denominator (i.e. x^3 is greater than x^2). In such a case, we have to divide first before we proceed. Let us carry out the division as follows:

$$\begin{array}{r}
x + 3 \\
x^2 - x - 2 \overline{\smash{)}\,x^3 + 2x^2 - x - 11} \\
\underline{x^3 - x^2 - 2x} \\
3x^2 + x - 11 \\
\underline{3x^2 - 3x - 6} \\
4x - 5
\end{array}$$

Hence, using the remainder which is $4x - 5$, the fraction in our question can be written as:
$$\frac{x^3 + 2x^2 - x - 11}{x^2 - x - 2} = x + 3 + \frac{4x - 5}{x^2 - x - 2}$$

We can now resolve $\frac{4x - 5}{x^2 - x - 2}$ into partial fractions. The denominator can be factorized to give:

$x^2 - x - 2 = (x - 2)(x + 1)$. Hence, the fraction to resolve can be written as $\frac{4x - 5}{(x - 2)(x + 1)}$

Therefore we resolve as follows:
$$\frac{4x - 5}{(x - 2)(x + 1)} = \frac{A}{x - 2} + \frac{B}{x + 1}$$
$$\frac{4x - 5}{(x - 2)(x + 1)} = \frac{A(x + 1) + B(x - 2)}{(x - 2)(x + 1)}$$

Equating the numerators gives:
$$4x - 5 \equiv A(x + 1) + B(x - 2)$$

Substitute $x = -1$ in order to get B. This gives:
$$4(-1) - 5 = A(-1 + 1) + B(-1 - 2)$$
$$-4 - 5 = A(0) + B(-3)$$
$$-9 = 0 - 3B$$
$$3B = 9$$
$$B = \frac{9}{3}$$
$$B = 3$$

Substitute $x = 2$ in order to find A. This gives:
$$4x - 5 \equiv A(x + 1) + B(x - 2)$$
$$4(2) - 5 = A(2 + 1) + B(2 - 2)$$
$$8 - 5 = A(3) + B(0)$$
$$3 = 3A + 0$$
$$A = \frac{3}{3}$$
$$A = 1$$

Hence A = 1 and B = 3

Therefore, $\frac{x^3 + 2x^2 - x - 11}{x^2 - x - 2} = x + 3 + \frac{1}{x - 2} + \frac{3}{x + 1}$

16. If $\frac{x^2 + x - 1}{(x + 1)(x - 1)} = A + \frac{B}{x + 1} + \frac{C}{x - 1}$ where A, B and C are constants, find A + 2B – C

Solution

If we expand the denominators, it gives:
$$(x + 1)(x - 1) = x^2 - x + x - 1$$
$$= x^2 - 1$$

This shows that the numerator and the denominator are of the same degree (i.e. both have the same highest power of x). Hence we have to divide first. This is as shown below.

$$\begin{array}{r} 1 \\ x^2-1\overline{)x^2+x-1} \\ \underline{x^2+0x-1} \\ x \end{array}$$ (Note that $0x$ is added since there is no term in x)

Therefore with x as the remainder, we now write the fraction as follows:

$$\frac{x^2+x-1}{(x+1)(x-1)} = 1 + \frac{x}{(x+1)(x-1)}$$ (This shows that A = 1)

Let us now resolve $\dfrac{x}{(x+1)(x-1)}$ into partial fractions as follows.

$$\frac{x}{(x+1)(x-1)} = \frac{B}{x+1} + \frac{C}{x-1}$$

$$\frac{x}{(x+1)(x-1)} = \frac{B(x-1)+C(x+1)}{(x+1)(x-1)}$$

Equating the numerators gives:

$x \equiv B(x-1) + C(x+1)$

If we put $x = 1$, we obtain C as follows:

1 = B(1 − 1) + C(1 + 1)
1 = B(0) + C(2)
1 = 0 + 2C
1 = 2C
$C = \dfrac{1}{2}$

If we put $x = -1$, we obtain B as follows:

−1 = B(−1 − 1) + C(−1 + 1)
−1 = B(−2) + C(0)
−1 = −2B + 0
2B = 1
$B = \dfrac{1}{2}$

Hence, $B = \dfrac{1}{2}$ and $C = \dfrac{1}{2}$. And recall that A = 1

Therefore, $A + 2B - C = 1 + 2(\dfrac{1}{2}) - \dfrac{1}{2}$

$$= 1 + 1 - \frac{1}{2}$$

$$= 2 - \frac{1}{2}$$

$$= 1\frac{1}{2}$$

17. If $\dfrac{3x^2-7}{x^3+2x^2-8x} \equiv \dfrac{7}{8x} + \dfrac{P}{x+4} + \dfrac{Q}{x-2}$, find P + Q

Solution

Looking at the denominators on both sides of the identity shows that the factors of $x^3 + 2x^2 - 8x$ are x, $(x + 4)$ and $(x - 2)$. This means that:

$x^3 + 2x^2 - 8x = x(x + 4)(x - 2)$

This shows that $\dfrac{7}{8x}$ should be written as $\dfrac{\frac{7}{8}}{x}$. If we use $\dfrac{7}{8x}$, then the denominators on both sides of the identity will not be identical. For the two denominators to be identical, we must write the question above as follows:

$$\dfrac{3x^2-7}{x^3+2x^2-8x} \equiv \dfrac{\frac{7}{8}}{x} + \dfrac{P}{x+4} + \dfrac{Q}{x-2}$$

We now continue our working as follows:

$$\dfrac{3x^2-7}{x(x+4)(x-2)} \equiv \dfrac{\frac{7}{8}(x+4)(x-2) + P(x)(x-2) + Q(x)(x+4)}{x(x+4)(x-2)}$$

Equating the numerators gives:

$3x^2 - 7 \equiv \dfrac{7}{8}(x+4)(x-2) + P(x)(x-2) + Q(x)(x+4)$

Substitute -4 for x to obtain P as follows

$3(-4)^2 - 7 \equiv \dfrac{7}{8}(-4+4)(-4-2) + P(-4)(-4-2) + Q(-4)(-4+4)$

$3(16) - 7 = \dfrac{7}{8}(0)(-6) + P(-4)(-6) + Q(-4)(0)$

$48 - 7 = 0 + 24P + 0$

$41 = 24P$

$P = \dfrac{41}{24}$

Let us put $x = 2$ in order to get Q.

$3x^2 - 7 \equiv \dfrac{7}{8}(x+4)(x-2) + P(x)(x-2) + Q(x)(x+4)$

$3(2)^2 - 7 \equiv \dfrac{7}{8}(2+4)(2-2) + P(2)(2-2) + Q(2)(2+4)$

$3(4) - 7 = \dfrac{7}{8}(6)(0) + P(2)(0) + Q(2)(6)$

$12 - 7 = 0 + 0 + 12Q$

$5 = 12Q$

$Q = \dfrac{5}{12}$

Therefore, $P = \dfrac{41}{24}$, and $Q = \dfrac{5}{12}$

Hence, $P + Q = \dfrac{41}{24} + \dfrac{5}{12}$

$$= \frac{41+10}{24}$$
$$= \frac{51}{24}$$
$$\therefore \quad P + Q = \frac{17}{8} \quad \text{(In its lowest term)}$$

18. Resolve $\dfrac{1 - 4x + 7x^2 - x^3 - x^4}{(x+3)(x^2+2)}$ into partial fractions.

Solutions

If we expand the denominator, it will give us a degree (highest power of x) of 3 (i.e. x^3). However the numerator contains x^4. Hence the fraction is an improper fraction. Therefore we have to divide the fraction before we proceed.

Let us expand the denominator as follows:
$$(x + 3)(x^2 + 2) = x^3 + 2x + 3x^2 + 6$$
$$= x^3 + 3x^2 + 2x + 6$$

We also have to rearrange the numerator so that the powers of x will be in descending order. This is necessary for easier division of the polynomial. Hence the fraction can be written as follows:
$$\frac{-x^4 - x^3 + 7x^2 - 4x + 1}{x^3 + 3x^2 + 2x + 6}$$

Let us now carry out the division as follows:

```
                       -x + 2
          _____
x³+3x²+2x+6) -x⁴ -  x³ + 7x² - 4x + 1
             -x⁴ - 3x³ - 2x²  - 6x
             _____
                   2x³ + 9x² + 2x + 1
                   2x³ + 6x² + 4x + 12
                   _____
                         3x² - 2x - 11
```

Note that we can no longer continue the division when there is no more term to bring down and the highest power of x in the remainder is smaller than that of the divisor. Hence, the fraction has been broken down to give:

$$\frac{1 - 4x + 7x^2 - x^3 - x^4}{(x+3)(x^2+2)} = 2 - x + \frac{3x^2 - 2x - 11}{(x+3)(x^2+2)} \quad \text{(Note that } -x + 2 \text{ has been written as } 2 - x\text{)}$$

Let us now resolve $\dfrac{3x^2 - 2x - 11}{(x+3)(x^2+2)}$ into partial fractions as follows:

$$\frac{3x^2 - 2x - 11}{(x+3)(x^2+2)} = \frac{A}{x+3} + \frac{Bx + C}{(x^2+2)}$$

$$\frac{3x^2 - 2x - 11}{(x+3)(x^2+2)} = \frac{A(x^2+2) + (Bx+C)(x+3)}{(x+3)(x^2+2)}$$

Equating the numerators gives:
$$3x^2 - 2x - 11 \equiv A(x^2 + 2) + (Bx + C)(x + 3)$$

Substitute -3 for x in order to find A. This gives:

$3(-3)^2 - 2(-3) - 11 = A((-3)^2 + 2) + (B(-3) + C)(-3 + 3)$

$3(9) + 6 - 11 = A(9 + 2) + (-3B + C)(0)$

$27 + 6 - 11 = 11A + 0$

$22 = 11A$

$A = \dfrac{22}{11}$

$A = 2$

Let us expand the brackets in the identity above.

$3x^2 - 2x - 11 \equiv A(x^2 + 2) + (Bx + C)(x + 3)$

$\equiv Ax^2 + 2A + Bx^2 + 3Bx + Cx + 3C$

$\equiv Ax^2 + Bx^2 + 3Bx + Cx + 2A + 3C$

$3x^2 - 2x - 11 \equiv (A + B)x^2 + (3B + C)x + 2A + 3C$

Comparing coefficients of like terms on both sides of the identity shows that:

$A + B = 3$ ……………Equation (1)

$3B + C = -2$ ……………Equation (2)

Since we have obtained the value of A as 2, let us substitute this in equation (1) as follows:

$A + B = 3$ ……………Equation (1)

$2 + B = 3$ (Note that A = 2 as obtained above)

$B = 3 - 2$

$B = 1$

From equation (2) we have that:

$3B + C = -2$ ……………Equation (2)

$3(1) + C = -2$

$3 + C = -2$

$C = -2 - 3$

$C = -5$

Hence, A = 2, B = 1 and C = −5

Therefore, $\dfrac{1 - 4x + 7x^2 - x^3 - x^4}{(x+3)(x^2+2)} = 2 - x + \dfrac{2}{x+3} + \dfrac{x-5}{(x^2+2)}$

Exercise 10

1. Express $\dfrac{5x - 1}{(x+3)(x-5)}$ in partial fractions

2. Resolve $\dfrac{2x + 11}{2x^2 - 3x - 9}$ into partial fractions.

3. Express $\dfrac{8}{x^2 - 16}$ in partial fractions.

4. Resolve $\dfrac{3x-10}{2x(x-5)}$ into partial fractions.

5. Resolve $\dfrac{x+7}{(x-2)^2}$ into partial fractions.

6. Express $\dfrac{6+4x-5x^2}{x^2(2x+3)}$ in partial fractions.

7. Resolve $\dfrac{2x^2-5x+11}{(x+1)(x-2)^2}$ into partial fractions.

8. Resolve $\dfrac{5x^2-x+2}{x(x^2+3)}$ into partial fractions.

9. Resolve $\dfrac{3x^2+2x+8}{(x^2+2)(x+1)}$ into partial fractions

10. Resolve $\dfrac{23x-31}{(x-2)(x^2+2x-3)}$ into partial fractions

11. Express $\dfrac{5x^2+2x-9}{(x+3)(x^2-3x-3)}$ in partial fractions

12. Resolve $\dfrac{x^2+x-7}{(x-1)^3}$ into partial fractions

13. Resolve $\dfrac{2x^3+3x^2-x-5}{x^2(x^2-1)}$ into partial fractions

14. Resolve $\dfrac{x^2-3x-1}{x^2-2x-3}$ into partial fractions.

15. Resolve $\dfrac{x^3+8x^2+13x-29}{x^2+3x-4}$ into partial fractions

16. If $\dfrac{5x^2+13x-14}{(x-1)(x+3)} = A + \dfrac{B}{x-1} + \dfrac{C}{x+3}$ where A, B and C are constants, find 2A + 5B – 7C

17. If $\dfrac{2x^2+5}{2x^3+14x^2+20x} \equiv \dfrac{1}{2x} + \dfrac{P}{x+5} + \dfrac{Q}{x+2}$, find the values of P and Q

18. Resolve $\dfrac{4-2x+5x^2+2x^3-x^4}{(x-1)(x^2+1)}$ into partial fractions.

CHAPTER 11
RADICAL EQUATIONS

Equations which consist of square root signs are called radical equations. In solving such equations, we keep squaring both sides of the equation until the square root sign is finally removed. Then we solve the resulting equation after the removal of the root sign.

However, it is important that we recall the following rules when simplifying surds:

1. $\sqrt{a} \times \sqrt{b} = \sqrt{ab}$
2. $\dfrac{\sqrt{a}}{\sqrt{b}} = \sqrt{\dfrac{a}{b}}$
3. $\sqrt{a} \times \sqrt{a} = a$

Examples

1. Solve the equation $\sqrt{x-2} = 3$

<u>Solution</u>

$\sqrt{x-2} = 3$

Square both sides in order to remove the root sign. This gives:

$(\sqrt{x-2})^2 = 3^2$

$x - 2 = 9$

Note that the expression $(\sqrt{x-2})^2$ gives $x - 2$, i.e. simply take the term in the root sign.

$x = 9 + 2$

$x = 11$

2. Solve $\sqrt{x+14} = \sqrt{6-x}$

<u>Solution</u>

$\sqrt{x+14} = \sqrt{6-x}$

Squaring both sides gives:

$(\sqrt{x+14})^2 = (\sqrt{6-x})^2$

$x + 14 = 6 - x$

$x + x = 6 - 14$

$2x = -8$

$x = \dfrac{-8}{2}$

$x = -4$

3. Solve the equation $\sqrt{x} + \sqrt{2x-2} = 7$

Solution

$\sqrt{x} + \sqrt{2x-2} = 7$

Square both sides of the equation in order to remove the root sign. This gives:

$(\sqrt{x} + \sqrt{2x-2})^2 = 7^2$

$(\sqrt{x} + \sqrt{2x-2})(\sqrt{x} + \sqrt{2x-2}) = 49$

In order to expand the left hand side, simply use each term in the first bracket to multiply the second bracket. This is as shown below.

$(\sqrt{x})(\sqrt{x} + \sqrt{2x-2}) + \sqrt{2x-2}(\sqrt{x} + \sqrt{2x-2}) = 49$

$x + \sqrt{x}\sqrt{2x-2} + \sqrt{x}\sqrt{2x-2} + 2x - 2 = 49$

$x + 2x - 2 + \sqrt{x(2x-2)} + \sqrt{x(2x-2)} = 49$

$3x - 2 + 2\sqrt{x(2x-2)} = 49$ [Note that $\sqrt{x(2x-2)} + \sqrt{x(2x-2)} = 2\sqrt{x(2x-2)}$]

$3x + 2\sqrt{2x^2 - 2x} = 49 + 2$

$3x + 2\sqrt{2x^2 - 2x} = 51$

$2\sqrt{2x^2 - 2x} = 51 - 3x$

Now that we have a single root sign, we have to square both sides of the equation to remove the root sign again. This gives:

$(2\sqrt{2x^2 - 2x})^2 = (51 - 3x)^2$

$(2)^2(\sqrt{2x^2 - 2x})^2 = (51 - 3x)(51 - 3x)$

$4(2x^2 - 2x) = 2601 - 153x - 153x + 9x^2$

$8x^2 - 8x = 2601 - 306x + 9x^2$

$0 = 9x^2 - 8x^2 + 8x - 306x + 2601$

$0 = x^2 - 298x + 2601$

Or, $x^2 - 298x + 2601$

Solving this equation by factorization gives:

$(x - 9)(x - 289) = 0$

$\therefore$ $x = 9$ or $x = 289$

After solving a radical equation it is necessary to check if the values obtained are actually the true solutions of the equation. Hence, let us substitute each of the values of x obtained above into the equation. The equation is:

$\sqrt{x} + \sqrt{2x-2} = 7$

When $x = 9$, the left hand side of the equation gives:

$\sqrt{9} + \sqrt{2(9) - 2} = 3 + \sqrt{18 - 2}$

$= 3 + \sqrt{16}$

$= 3 + 4$

$= 7$

Since 7 is also the value on the right hand side of the equation, then $x = 9$ is a solution of the equation.

Recall that the equation is $\sqrt{x} + \sqrt{2x - 2} = 7$
When $x = 289$, the left hand side of the equation gives:
$$\sqrt{289} + \sqrt{2(289) - 2} = 17 + \sqrt{578 - 2}$$
$$= 17 + \sqrt{576}$$
$$= 17 + 24$$
$$= 41$$
Since 41 is not what we have on the right hand side of the equation, then $x = 289$ is not a solution of the equation.
Therefore, the solution of the equation is $x = 9$
Note that 289 which is not a solution of the equation is called an extraneous root of the equation.

4. Solve the equation $\sqrt{3x + 13} - x = 1$

Solution

$\sqrt{3x + 13} - x = 1$

Collect terms containing the root sign (radical) on one side of the equation, and move other terms to the other side of the equation. This means we have to isolate the radical term. This gives:

$\sqrt{3x + 13} = 1 + x$

We can now square both sides of the equation as follows:

$(\sqrt{3x + 13})^2 = (1 + x)^2$
$3x + 13 = (1 + x)(1 + x)$
$3x + 13 = 1 + x + x + x^2$
$3x + 13 = x^2 + 2x + 1$
$0 = x^2 + 2x - 3x + 1 - 13$
$0 = x^2 - x - 12$
$x^2 - x - 12 = 0$

Solving this equation by factorization gives:

$(x - 4)(x + 3) = 0$

Hence, $x = 4$ or $x = -3$

Let us check the values of x obtained in order to know if they are solutions of the equation. The equation is:

$\sqrt{3x + 13} - x = 1$

When $x = 4$, the left hand side of the equation gives:

$\sqrt{3x + 13} - x = \sqrt{3(4) + 13} - 4$
$= \sqrt{12 + 13} - 4$
$= \sqrt{25} - 4$
$= 5 - 4 = 1$

Since 1 is also the value on the right hand side of the equation, the $x = 4$ is a solution of the equation.

When $x = -3$, the left hand side of the equation gives:
$$\sqrt{3x + 13} - x = \sqrt{3(-3) + 13} - (-3)$$
$$= \sqrt{-9 + 13} + 3$$
$$= \sqrt{4} + 3$$
$$= 2 + 3$$
$$= 5$$
Since 5 is not what we have on the right hand side of the equation, then $x = -3$ is not a solution of the equation.
Therefore, the solution of the equation is $x = 4$

5. Solve for x if $\sqrt{x} + 2\sqrt{x + 8} = 7$
Solution
$\sqrt{x} + 2\sqrt{x + 8} = 7$
Squaring both sides of the equation gives:
$$(\sqrt{x} + 2\sqrt{x + 8})^2 = 7^2$$
$$(\sqrt{x} + 2\sqrt{x + 8})(\sqrt{x} + 2\sqrt{x + 8}) = 49$$
$$x + 2\sqrt{x(x + 8)} + 2\sqrt{x(x + 8)} + 4(x + 8) = 49$$
$$x + 4\sqrt{x(x + 8)} + 4x + 32 = 49$$
$$4\sqrt{x^2 + 8x} = 49 - 32 - 4x - x$$
$$4\sqrt{x^2 + 8x} = 17 - 5x$$
Squaring both sides again gives:
$$(4\sqrt{x^2 + 8x})^2 = (17 - 5x)^2$$
$$4^2(\sqrt{x^2 + 8x})^2 = (17 - 5x)(17 - 5x)$$
$$16(x^2 + 8x) = 289 - 85x - 85x + 25x^2$$
$$16x^2 + 128x = 289 - 170x + 25x^2$$
$$0 = 25x^2 - 16x^2 - 170x - 128x + 289$$
$$0 = 9x^2 - 298x + 289$$
$$9x^2 - 298x + 289 = 0$$
Let us solve this equation by factorization (Note that quadratic formula can also be used) as follows:
$$9x^2 - 9x - 289x + 289 = 0$$
$$9x(x - 1) - 289(x - 1) = 0$$
$$(x - 1)(9x - 289) = 0$$
Hence, $x = 1$ or $x = \dfrac{289}{9}$

$\dfrac{289}{9}$ cannot be the solution to the equation since it is a fraction and it is too large (the right hand side is just 7).

Therefore the solution to the equation is $x = 1$

6. Solve for x if $\sqrt{x+7} + 3x = 9$

Solution

$\sqrt{x+7} + 3x = 9$

Collect term containing the radical on the left hand side of the equation. This gives:

$\sqrt{x+7} = 9 - 3x$

Square both sides of the equation

$(\sqrt{x+7})^2 = (9-3x)^2$

$x + 7 = (9-3x)(9-3x)$

$x + 7 = 81 - 27x - 27x + 9x^2$

$0 = 81 - 54x + 9x^2 - x - 7$

$0 = 9x^2 - 55x + 74$

$9x^2 - 55x + 74 = 0$

Let us solve this equation by using quadratic formula. From the equation:

$a = 9$, $b = -55$ and $c = 74$

$x = \dfrac{-b \pm \sqrt{b^2 - 4ac}}{2a}$

$= \dfrac{-(-55) \pm \sqrt{(-55)^2 - (4 \times 9 \times 74)}}{2 \times 9}$

$= \dfrac{55 \pm \sqrt{3025 - 2664}}{18}$

$= \dfrac{55 \pm \sqrt{361}}{18}$

$= \dfrac{55 \pm 19}{18}$

$x = \dfrac{55 + 19}{18}$ or $x = \dfrac{55 - 19}{18}$

$x = \dfrac{74}{18}$ or $x = \dfrac{36}{18}$

$x = \dfrac{37}{9}$ or $x = 2$

$\dfrac{37}{9}$ cannot be the solution of the equation if substituted into the equation.

Therefore the solution to the equation is $x = 2$

7. Find the value of x in the equation $\sqrt{x+1} + \sqrt{x+6} = 5$

Solution

$\sqrt{x+1} + \sqrt{x+6} = 5$

Squaring both sides of the equation gives:

$(\sqrt{x+1} + \sqrt{x+6})^2 = 5^2$

$(\sqrt{x+1} + \sqrt{x+6})(\sqrt{x+1} + \sqrt{x+6}) = 25$

$(\sqrt{x+1} + \sqrt{x+6})(\sqrt{x+1} + \sqrt{x+6}) = 25$

$x + 1 + \sqrt{(x+1)(x+6)} + \sqrt{(x+1)(x+6)} + x + 6 = 25$

$x + 1 + 2\sqrt{(x+1)(x+6)} + x + 6 = 25$

$2x + 7 + 2\sqrt{(x+1)(x+6)} = 25$

Isolating the radical term on the left hand side gives:

$2\sqrt{(x+1)(x+6)} = 25 - 2x - 7$

$2\sqrt{x^2 + 7x + 6} = 18 - 2x$

Divide both sides by 2. This gives:

$\sqrt{x^2 + 7x + 6} = 9 - x$

Squaring both sides of the equation gives:

$(\sqrt{x^2 + 7x + 6})^2 = (9 - x)^2$

$x^2 + 7x + 6 = (9 - x)(9 - x)$

$x^2 + 7x + 6 = 81 - 18x + x^2$

$x^2 - x^2 + 7x + 18x = 81 - 6$

$25x = 75$

$x = \dfrac{75}{25}$

$x = 3$

8. Find the value of x if $\sqrt{2x^2 + 7} = 3$

Solution

$\sqrt{2x^2 + 7} = 3$

Squaring both sides of the equation gives:

$(\sqrt{2x^2 + 7})^2 = 3^2$

$2x^2 + 7 = 9$

$2x^2 = 9 - 7$

$2x^2 = 2$

$x^2 = 1$

$x = \sqrt{1}$

$x = \pm 1$

$x = 1$ or $x = -1$

Let us check if the two values of x are solutions of the equation. The equation is:

$\sqrt{2x^2 + 7} = 3$

When $x = 1$, the left hand side of the equation gives:

$\sqrt{2x^2 + 7} = \sqrt{2(1)^2 + 7}$

$= \sqrt{2 + 7}$

$= \sqrt{9}$

$= 3$

Hence $x = 1$ is a solution of the equation.

When $x = -1$, the left hand side of the equation gives:

$\sqrt{2x^2 + 7} = \sqrt{2(-1)^2 + 7}$

$= \sqrt{2(1) + 7}$

$= \sqrt{2 + 7}$

$= \sqrt{9} = 3$

Hence $x = -1$ is a solution of the equation.

Therefore both $x = 1$, and $x = -1$ are solutions of the equation.

9. Solve the equation $\sqrt{5x^2 - 2x} = 4$

Solution

$\sqrt{5x^2 - 2x} = 4$

Squaring both sides of the equation gives:

$(\sqrt{5x^2 - 2x})^2 = 4^2$

$5x^2 - 2x = 16$

$5x^2 - 2x - 16 = 0$

Solving this equation by factorization gives:

$5x^2 - 10x + 8x - 16 = 0$

$5x(x - 2) + 8(x - 2) = 0$

$(x - 2)(5x + 8) = 0$

$x = 2$ or $x = -\dfrac{8}{5}$

If we substitute $x = 2$ and $x = -\dfrac{8}{5}$ into the equation, we will find out that the two values of x are solutions of the equation.

Therefore, both $x = 2$ and $x = -\dfrac{8}{5}$ are solutions of the equation.

10. Solve $\sqrt{x^2 - 9} - x = -1$

Solution

Take $-x$ to the other side of the equation in order to isolate the radical term. This gives:

$\sqrt{x^2 - 9} = x - 1$

Squaring both sides of the equation gives:

$(\sqrt{x^2 - 9})^2 = (x - 1)^2$

$x^2 - 9 = x^2 - 2x + 1$

$x^2 - x^2 + 2x = 1 + 9$

$2x = 10$

$$x = \frac{10}{2}$$
$$x = 5$$

Exercise 11

1. Solve the equation $\sqrt{2x-3} = 7$
2. Solve $\sqrt{2x-9} = \sqrt{x-2}$
3. Solve the equation $\sqrt{2x} - \sqrt{4x-2} = -1$
4. Solve the equation $\sqrt{2x+16} - x = -4$
5. Solve for x if $\sqrt{4x} + 5\sqrt{x+3} = 12$
6. Solve for x if $\sqrt{2x-6} + 3x = 17$
7. Find the value of x in the equation $\sqrt{x-1} + \sqrt{x+4} = 5$
8. Find the value of x if $\sqrt{3x^2 + 13} = 11$
9. Solve the equation $\sqrt{2x^2 - 4x} = 4$
10. Solve $\sqrt{5x^2 + 4} - 2x = 1$

SOLUTION TO EXERCISES

Exercise 1

1. $\dfrac{m+d}{m+f}$ 2. $\dfrac{x-7}{x-4}$ 3. $\dfrac{2-x}{3x-5}$ 4. $-\dfrac{(4m+n)}{(n+m)}$ 5. $\dfrac{p-r}{p+r}$ 6. $\dfrac{6a+7}{3b}$ 7. $\dfrac{13}{5(2x-y)}$ 8. $\dfrac{1+9m}{6mn}$

9. $\dfrac{5a^2+13a+4}{(a-2)(a+3)}$ 10. $\dfrac{3ab^2+4b^2-3a^3}{12a^2b^3}$ 11. $\dfrac{4+5n}{(3m-2n)(3m+2n)}$

Exercise 2

1. $m=-4$ 2. $b=-\dfrac{3}{2}$ or 6 3. $x=7$ or 4.875 4. $x=-8$ 5. $x=\dfrac{7}{3}$ 6 (a) $a=-\dfrac{12}{5}$

(b) $a=6$ or -5 7. (a) $m=-2$ (b) $m=-\dfrac{9}{5}$ 8 (a) $x=\dfrac{2}{5}$ or -3 (b) $x=-\dfrac{5}{2}$ or 3 9. $m=\dfrac{22}{5}$

10. $x=-\dfrac{41}{113}$ 11. 4 12. $\dfrac{33}{26}$

Exercise 3

1. $x=1, y=2$ 2. $a=\dfrac{1}{4}, b=-1$ 3. $c=2, d=3$ 4. $a=\dfrac{4}{3}, b=\dfrac{2}{3}$ 5. $p=-\dfrac{1}{2}, q=\dfrac{1}{2}$

6. $c=2, d=-5$ 7. $x=\dfrac{3}{10}, y=\dfrac{2}{5}$ 8. $m=\dfrac{9}{17}, n=9$

Exercise 4

1. $x=9$ or -9 2. (a) $x=2$ or $-\dfrac{24}{5}$ (b) $x=\dfrac{7}{3}$ or -3 (c) $x=-\dfrac{3}{4}$ or $\dfrac{33}{4}$ 3 (a) $x=\dfrac{5}{4}$ or -3

(b) No solution 4 (a) $x=\dfrac{7}{2}$ or -2 (b) $x=-5$ or -13 5 (a) $x=2$ (b) $x=35$ or -3

(c) $x=-\dfrac{1}{2}$, or $\dfrac{1}{2}$ 6 (a) $x=2$ or -9 (b) $x=0, \dfrac{9}{2}, x=\dfrac{1}{2}$ or 4 7 (a) $x=12$ (b) $x=1$

8 (a) No solution (b) $x=5$ or -2 9 (a) $x=-\dfrac{2}{3}$ or 3 (b) $x=3.63, -17.63, 4$ or -18

10 (a) $x=\dfrac{3}{5}$ or 2 (b) $x=1.30, -5.80, 0.84$ or -5.34

Exercise 5

1. $x\leq -2$ or $x\geq \dfrac{4}{5}$ 2. $-5<x<\dfrac{19}{3}$ 3. $-3\leq x\leq 19$ 4. $m<-1$ or $m>-\dfrac{3}{7}$ 5. $x<-3$ or $x>5$

6. $-3<m<2$ 7. $2<m\leq 5$ 8. $-2<x\leq -1$ 9. $x<-3$ or $x>3$ 10. $y<6$ or $y>-6$

11. $a\leq -2.1$ or $a\geq 5.4$ 12. $y\leq -3$ or $y\geq 3$ 13. $m\leq -1$ or $m\geq 1$ 14. $x>-\dfrac{1}{2}$ or $x\leq 2$

15. $-4\leq y\leq 4$

Exercise 6

1. $x=\dfrac{16}{5}$ 2. $x=\dfrac{13}{11}$ 3. $x=\dfrac{14}{13}$ 4. $x=0.42$ 5. $x=-41.39$ 6. $x=0.474$ 7. $x=2.50$ or -0.34

8. $x = 4.14$ or 0.36 9. $x = 3$ 10. $x = 2$ 11. $x = 0$ 12. $x = 1, y = -1$ 13. $x = 5, y = \dfrac{20}{3}$
14. $y = -\dfrac{4}{9}, x = \dfrac{19}{9}$ 15. $x = -\dfrac{4}{3}$ or 2

Execise 7

1 (a) $\dfrac{8}{3}$ (b) $\dfrac{5}{3}$ (c) $\dfrac{34}{9}$ (d) $\dfrac{34}{15}$ (e) $\dfrac{8}{15}$ (f) $\dfrac{34}{25}$ (g) $\dfrac{152}{27}$ 2. $4x^2 - 188x + 121 = 0$
3. $40x^2 + 89x + 40 = 0$ 4. $2x^2 + 15x - 1 = 0$ 5. $49x^2 - 189x + 9 = 0$ 6. $8x^2 + 305x - 216 = 0$
7 (a) $\dfrac{\sqrt{24}}{5}$ (b) $\dfrac{8\sqrt{24}}{25}$ 8. $5\dfrac{9}{20}$ 9. $\dfrac{13\sqrt{13}}{8}$ 10 (a) $3x^2 - 28x + 32 = 0$ (b) $3x^2 - 8x - 16 = 0$
11. $m = -9$ and $n = \dfrac{2}{3}$ or $m = -2$ and $n = 3$ 12. $P = 1$ or -23 13. $k = -\dfrac{28}{3}$ 14. $\dfrac{84}{5}$ 15 (a) $y = -\dfrac{217}{8}$
(b) $x = -\dfrac{11}{4}$ 16 (a) $y = \dfrac{513}{16}$ (b) $x = -\dfrac{15}{8}$ 17 (a) 1 and 3 or -1 and -3 (b) $k = -6$ or 10
18. $3\sqrt{3}$ 19. $K = -8$ 20. $\dfrac{\sqrt{5}}{2}$ and $\sqrt{5}$ or $-\dfrac{\sqrt{5}}{2}$ and $-\sqrt{5}$

Exercise 8

1 (a) It is not a function   (b) It is not a function   (c) It is not a function   (d) It is a function
2 (a) -5 (b) 22   (c) -1.625 (d) $81x^3 - 162x^2 + 108x - 26$ (e) $3x^3 + 27x^2 + 81x - 79$
3 (a) $-\dfrac{13}{5}$ (b) $p = \dfrac{7}{4}$ (c) $-\dfrac{153}{5}$ 4 (a) $3x - 2$ (b) $16 - 3x$ (c) $2x^2 - 5x - 3$ (d) $-\dfrac{4}{5}$
(e) $h(x) = x - 16$ (f) -14 5 (a) $2x - 4$ (b) $2x - 2$ 6 (a) $18(4x^2 + 4x + 1)$ (b) $18x^2$
(c) $36x^2 + 1$ 7 (a) Even   (b) Odd   8 (a) Even   (b) Even   (c) Neither   (d) Even
9 (a) $\dfrac{x+1}{5}$ (b) $\sqrt[3]{\dfrac{x-2}{9}}$ (c) $\sqrt[5]{\dfrac{x}{2}}$ 10 (a) $2\sqrt[3]{3} + 1$ (b) $\dfrac{1 + 2x^2}{5x^2 - 2}$ 11 (a) $\dfrac{x+1}{5 - 2x}$ (b) 0
12 (a) $2x^2 + 3x - 9$ (b) -7 13. $\dfrac{2x - 2}{5} + 7$ 14. 47   15 (a) $2x - 5$ (b) $\dfrac{x+5}{2}$ (c) -9
(d) 2   (e) $4x - 7$ 16 (a) $7x - 12$ (b) $14x^2 + 23$

Exercise 9

1 (a) $5x^2 + 5x + 6$ (b) $-2x^2 + 4x - 22$ (c) $3x^3 - 8x^2 + 6x - 11$ (d) $3x^3 + 7x^2 - 4x + 21$
(e) $9x^3 + 4x^2 + 2x + 9$ 2 (a) -82 (b) -10 3. $10x^6 + 11x^5 - 29x^4 - 2x^3 - x^2 - 2x$
4. $6x^6 - 11x^5 + 25x^4 - 5x^3 + 19x^2 - 14x - 20$ 5. $2x^6 - 3x^5 + 6x^4 - 19x^3 + 15x^2 - 30x + 45$
6. $x - 2$ 7. $2x^2 - 5x + 8$ 8. $3x^2 - 5$ 9. $2x^2 - xy + 3y^2$ 10. $5x^2 - 2xy - y^2$
11. $2x^2 + 3x + 27$ remainder 126   12. $5x^2 - 8x + 5$ remainder -9 13 (a) $x = \dfrac{15}{2}$ and 2
(b) $x = -\dfrac{3}{2}$ and $\dfrac{1}{2}$ 14. $(x+1)(x-2)(x-3)$. Hence, $x = -1, 2$ or 3 15. $x = 1, -2, 3$ or $-\dfrac{5}{2}$ 16. 11
17. $7\dfrac{1}{4}$ 18. $2x + 3$ and $2x + 1$ 19. $x = 0, -\dfrac{1}{2}$ or $\dfrac{1}{2}$ 20. $3x^2 - 5$ 21. $a = 2, b = 8$
22 (a) $m = -9, n = 10$ (b) $x^2 - x - 10$ (c) $x = 1, 3.7$ or -2.7 23. $m = \dfrac{1}{2}, n = 39$

Exercise 10

1. $\dfrac{2}{x+3}+\dfrac{3}{x-5}$ 2. $\dfrac{17}{9(x-3)}-\dfrac{16}{9(2x+3)}$ 3. $\dfrac{1}{x-4}+\dfrac{1}{x+4}$ 4. $\dfrac{1}{x}+\dfrac{1}{2(x-5)}$ 5. $\dfrac{1}{x-2}+\dfrac{9}{(x-2)^2}$

6. $\dfrac{2}{x^2}-\dfrac{5}{2x+3}$ 7. $\dfrac{2}{x+1}+\dfrac{3}{(x-2)^2}$ 8. $\dfrac{2}{3x}+\dfrac{13x-3}{3(x^2+3)}$ 9. $\dfrac{2}{x^2+2}+\dfrac{3}{x+1}$

10. $\dfrac{3}{x-2}-\dfrac{5}{x+3}+\dfrac{2}{x-1}$ or $\dfrac{3}{x-2}+\dfrac{11-3x}{x^2+2x-3}$ 11. $\dfrac{2}{x+3}+\dfrac{3x-1}{x^2-3x-3}$

12. $\dfrac{1}{x-1}+\dfrac{3}{(x-1)^2}-\dfrac{5}{(x-1)^3}$ 13. $\dfrac{1}{x}+\dfrac{5}{x^2}+\dfrac{x-2}{x^2-1}$ 14. $1-\dfrac{1}{4(x-3)}-\dfrac{3}{4(x+1)}$

15. $x+5+\dfrac{17}{5(x+4)}-\dfrac{7}{5(x-1)}$ 16. 1 17. $P=-\dfrac{11}{6},\ Q=-\dfrac{13}{12}$ 18. $1-x+\dfrac{4}{x-1}+\dfrac{3x-1}{x^2+1}$

Exercise 11

1. $x=26$ 2. $x=7$ 3. $x=4\dfrac{1}{2}$ or $\dfrac{1}{2}$ 4. $x=10$ 5. $x=1$ 6. $x=5$ 7. $x=5$ 8. $x=6$ or -6
9. $x=4$ or -2 10. $x=1$ or 3

If you have any enquiries, suggestions or information concerning this book, please contact the author through the email below.

Kingsley Augustine

kingzohb2@yahoo.com

Twitter handle: @kingzohb2

www.ingramcontent.com/pod-product-compliance
Lightning Source LLC
Chambersburg PA
CBHW080544220526
45466CB00010B/3028